Contents

Introduction

There is a well-known and very apt old adage that goes *'If you don't know how well you are doing, how do you know you are doing well?'* In non-construction sectors, this critical need to know how well you are doing is the primary driver for the use of performance measurement and performance self-assessment as the mechanisms by which firms accurately inform themselves of their true performance in every aspect of their business, including the performance of the firms that make up their supply chain. They then use the information to accurately compare their performance and that of their suppliers with the performance of the market leaders and with the expectations of the potential customers of their products.

Most importantly, regular performance measurement is also the means by which firms in other sectors detect unnecessary costs in the effective utilisation of labour and materials throughout their supply chain. They, and the firms in their supply chain, then use this accurate breakdown of unnecessary costs to work together to minimise them and convert the savings into higher profits and lower prices and thus make them more competitive.

There is no reason why the construction industry should not use similar techniques to reveal the unnecessary costs within the design and construction process and convert the savings into higher profits and lower prices. The 1994 Latham Report warned that unnecessary costs caused by the inefficient utilisation of labour and materials could be as high as 30% of the total cost of construction.

Subsequent research by the Building Services Research and Information Association (BSRIA), the Building Research Establishment (BRE) and the Construction Best Practice Programme (CBPP) has fully validated this 30% figure, which amounts to around £17 billion a year since the total annual UK construction expenditure is around £58 billion. Consequently, the construction industry has much to gain from importing accurate performance measurement techniques and using them to convert unnecessary costs into far higher profits and far lower prices.

In 2002, the UK Strategic Forum for Construction report *Accelerating Change* called for the UK construction industry to adopt best practice in performance measurement and continuous improvement from other sectors by introducing the following into their way of doing business:

❑ *'A culture of continuous improvement based on performance measurement.*
❑ *Consistent and continuously improving performance, and improved profitability, making it highly valued by its stakeholders.'*

This book is a companion volume to the author's earlier book *Building Down Barriers – A guide to construction best practice* (published by Taylor and Francis) that explained what improvements in performance, structure and culture end-users wanted from the UK construction industry in order to give far better value for money.

It builds on the general guidance given in the earlier book by focusing in depth on the unbreakable link between the elimination of unnecessary costs and the measurement of performance. It is intended to help those firms that are intent on adopting the recommendations from *Accelerating Change* by explaining why chief executives, executive board members and senior managers need to measure the true performance of the processes within their own business, and those within their suppliers' businesses. It explains why accurate performance measurement is fundamental to any drive for continued improvement and why assumptions and anecdotal evidence are inevitably a dangerous and destructive basis for any improvement programme and could damage long-term competitiveness.

The book explains why benchmarking will fail unless the firms involved in the benchmarking process use common performance measurement systems. It links its guidance with that provided in *The Construction Performance Drivers: A health check for your business* published by BQC Performance Management Ltd and endorsed by the British Quality Foundation. It explains how measurement systems, such as the BRE CALIBRE productivity measurement system, can be used at project level to detect and reduce the inefficient utilisation of labour and materials. It also explains how self-assessment at operative level can be equally effective at detecting and exposing inefficiencies, if done within the security of long-term strategic supply chain partnerships and with the help of an experienced facilitator.

The book is written in straightforward language, and in a format that ensures ease of use in each sector of the construction industry and in each management tier within each sector. It looks back over the numerous reports that have been produced in the past 50 years or so and shows that end user demands for the elimination of unnecessary costs have remained surprisingly constant. It looks at the barriers that have caused the industry's firms to continue to fail to meet the end users' expectations during that time. It simplifies and explains the key demands for improved performance made of the industry by the Latham and the Egan reports and links them to the key themes of the National Audit Office report *Modernising Construction* and the Confederation of

Construction Clients *Charter Handbook*. It explains how the key themes from the *Charter Handbook* have become the key themes in the latest Egan report *Accelerating Change*.

The book also explains why the end-user customers of constructed products should establish a level playing field by basing their selection of the design and construction team on the factual evidence of improved performance that has come from measurement.

Acknowledgements

The author wishes to acknowledge the help, advice and case history evidence he received from the members of the British Quality Foundation Construction Group, without which the book would have been considerably less relevant to the needs of those at the sharp end.

He also wishes to record his thanks to Taylor and Francis, the publishers of his first best practice guidebook *Building Down Barriers – A Guide to Construction Best Practice*. The Taylor and Francis book provided the essential background material that was refined and further developed to give a sound and common foundation to this book. This ensured that the guidance contained in both books was fully co-ordinated, based on the same historical foundations and used the same six goals of construction best practice, and thus avoided any risk of conflict where both books were being used to drive forward the different aspects of improved performance in the same organisation.

He also wishes to record his thanks to BSRIA, who provided additional case history advice and evidence from the research they had carried out subsequent to the publication of their Technical Note TN14/97 *Improving M & E Site Productivity* to assess the take up of performance measurement within the building services sector.

Last, but by no means least, he would like to record his thanks to the CALIBRE team at the Building Research Establishment for their advice and input into the book from their experiences of the application of the CALIBRE productivity measurement system.

1   Why Measure Anything?

From all sectors of the construction industry, from the client's professional representatives through the consultants and construction contractors to the specialist suppliers (sub-contractors), I am constantly being asked why the traditional design and construction process needs to be radically changed and improved. Equally constantly, my concern about the current and historic high levels of unnecessary costs are disbelieved, with the typical reaction being the one I received in early 2003 from the Chief Executive of a major specialist supplier.

> **Case history – Industry awards submissions**
> I had voiced concern after a construction industry awards assessment meeting about the consequences of the inefficient utilisation of labour and the fact that not one of the candidates had made mention of any improvement in their effective utilisation of labour and materials, or had made mention of the benefit of measuring their performance. A Chief Executive from a specialist supplier firm that was also on the judges panel said that the attitude of the candidates was entirely reasonable since efficiency levels were universally high across the industry.
>
> He went on to insist that in the case of his own firm, his operatives were achieving (and had always been achieving)

near maximum efficiency levels on all the sites on which they were involved. When I queried how he could be sure of his efficiency levels, his response was that there was never anyone in the yard therefore every one of his operatives must be effectively and efficiently utilised on the various sites all of the time.

We then had a lengthy discussion about the value of accurate and objective measurement of on-site performance and the very real and tangible commercial benefit that would automatically come from using the result of accurate performance measurement to drive out unnecessary costs and thus to drive up profits and drive down prices.

He vehemently insisted that the cost of such performance measurement could never be offset against cost savings because it was self-evident to him, as the Chief Executive, that efficiency levels in his firm could not be significantly improved.

Chief Executives and senior managers of construction contractors and specialist suppliers have echoed his views time and time again across the industry. They have also been echoed by some of the senior officials within the Department of Trade and Industry and by many in key positions in major industry bodies. The notable exceptions have been the National Audit Office (their report *Modernising Construction* (see Further reading) makes their position on the primary importance of measuring performance very clear) and the Confederation of Construction Clients in their *Charter Handbook* (see Further reading). Rethinking Construction has also recognised the critical importance of performance measurement and has made its position clear in various publications. All three organisations insist that the start of any improvement process must be the objective measurement of current performance.

As a senior representative of the DTI said *'If you don't know how well you are doing, how do you know you are doing well?'*

Unfortunately, the vast majority of industry firms appear not to be picking up this message about the importance of measuring current performance and I have yet to come across a firm that is regularly measuring its effective utilisation of labour and materials. I suspect the reason why very few seem to be paying attention to the message is that it is never directly linked with a message about the high levels of inefficiency and waste in the utilisation of labour and materials, and the concomitant high levels of unnecessary costs caused by the inefficiency and waste. Since no-one spells out the very considerable magnitude of the unnecessary costs that the investment in performance measurement would eradicate, it is not surprising that very few firms can see the point of performance measurement.

I rarely hear a senior industry figure from a key organisation, such as Rethinking Construction or the DTI, lay on the line in blunt and unambiguous terms the truth about current and historic low levels of effectiveness in the utilisation of labour and materials and then spell out the high levels of unnecessary costs that are directly caused by the ineffective utilisation of labour and materials. In fact, the concept of unnecessary costs and their cause seems to be totally beyond the understanding of the industry, even though it is well understood in other sectors where the primary driver of improvement in every successful firm is the continuous search for unnecessary costs in the utilisation of labour and materials down through the supply chain.

In other sectors, competitive success and higher profits come from all those in the supply chain collaborating to constantly measure their performance in the utilisation of labour and materials and then working together to devise ways of eliminating the causes of inefficiency so that unnecessary costs are continuously driven down.

So why should it be necessary to change the construction industry's attitude to performance measurement? Why should firms divert scarce money and resources away from traditional activities and invest in measuring their effective utilisation of labour and materials? Why should Chief Executives risk the adverse competitive consequences that would almost certainly flow from being honest about their current performance levels? Why should Rethinking Construction and the DTI embarrass the industry by focusing strongly on the high levels of unnecessary costs caused by the ineffective utilisation of labour and materials? Why should Rethinking Construction and the DTI insist that firms could only start to deliver real improvements if they are honest about their current levels of unnecessary costs and use performance measurement to reveal the truth? More specifically, why do I believe that performance measurement is an absolute imperative if improvement is to deliver any real, tangible and durable benefits to both the industry and its end-user customers?

The simple answer is that the only way that prices to the end user can be seriously reduced, the only way that the industry's profit margins can be seriously raised, the only way that the out-turn cost can always be held within the end user's initial budget, is by the elimination of the unnecessary costs that are caused by the ineffective utilisation of labour and materials. These unnecessary costs can only be eliminated if performance measurement is used to locate the true causes of the ineffective utilisation of labour and materials. Without performance measurement pointing the way, any claimed improvement process is merely tinkering with the edges and is unlikely to deliver any real and lasting benefits.

The investment in performance measurement will be rapidly repaid by the reduction in unnecessary costs that will come from the efficiency gains in utilisation of labour and materials in the supply chain firms. Objective performance measurement is the only way of locating the precise causes of the ineffective utilisation of labour and materials and unless the precise causes are exposed they cannot be eliminated. Evidence from the two Building Down Barriers pilot projects and from projects that have used the Building Research Establishment CALIBRE system, is that performance measurement will more than pay for itself within a single construction project.

The reason for the rapid payback is that the current low levels of effective utilisation of labour and materials make it easy to find big efficiency gains once performance measurement has revealed the truth about what is really happening on site. If your current level of effectiveness in the utilisation of labour and materials is around the 30% to 40% mark, it is comparatively easy to drive it up to the 50% to 60% mark if the design and construction firms in the supply chain work together in close collaboration to attack the causes of inefficiency revealed by performance measurement. It will obviously get a lot harder as efficiency levels near 100%, but the industry has a long way to go before that becomes a problem.

There is a major barrier to the realisation of the efficiency gains that ought to flow from the truth revealed by performance measurement. They will be impossible to realise if the design and construction supply chain team remain fragmented because there will be no way that the team could collaborate closely enough to eliminate the cross-firm causes of inefficiency.

The entire design and construction supply chain team *must* work together before any real progress can be made in the effective utilisation of labour and materials. The specialist supplier alone cannot reduce the disruption and reworking caused by errors and deficiencies in the drawings. This can only be done if the specialist supplier works in close collaboration with the design consultants within the mutually supportive and trusting relationship created by a long-term, stable, strategic supply chain partnering arrangement. Similarly, the specialist supplier alone cannot reduce the disruption and reworking caused by poor pre-planning of construction activities. This can only be done if the specialist supplier works in close collaboration with the construction contractor within the mutually supportive and trusting relationship created by a long-term, stable, strategic supply chain partnering arrangement.

These long-term, stable, strategic partnering relationships between the members of the design and construction supply chain are not beyond the wit of construction industry firms. *But* they do require people to set aside their traditional ways of thinking and working, they do require people to think laterally about different ways of working together, and they do require people to realise that such partnering relationships can be made to work at a strategic level that overarches individual projects. They also require people to realise that strategic supply chain partnering can be made to work without the need for a client to insist within the contract that such relationships must be in place before the contract can be signed.

None of these things are rocket science, all are relatively easy to achieve and all have been done for many years in other sectors where best practice advice and benchmarks are readily available.

Performance measurement is not just an optional bolt-on to the improvement process that can be included if funds permit. Performance measurement is the essential first stage to any improvement process that is intended to deliver real, long-lasting and tangible benefits in the form of significantly lower prices to the end user and significantly higher profit margins to industry firms, whilst enhancing the whole-life quality of the constructed product. Without performance measurement revealing the truth about how well it is doing, a firm will never be motivated to attack unnecessary costs. If a firm doesn't know how well it is doing, how can it decide what aspects of its performance are in most need of improvement?

The introduction of performance measurement enables Chief Executives to fully understand what is really happening at site level and to fully understand what is really hindering their effective utilisation of labour and materials. Performance measurement stops Chief Executives basing their improvement strategy on false assumptions about their effective utilisation of labour and materials. Performance measurement enables the improvement process to deliver real and measurable efficiency gains and ensures that the gains can be captured and converted into lower prices and higher profits.

Performance measurement provides Chief Executives and senior managers with accurate and undeniable evidence from sites that enables them to target and resolve the true causes of disruption and reworking that operatives have to face on every project time after time after time. Performance measurement puts an end to endless, fruitless and subjective arguments between site personnel and senior managers, and between the various members of the design and construction supply chain, about the magnitude and the causes of disruption and reworking on site.

Performance measurement provides incontrovertible evidence of the disruption and reworking on site that forces Chief Executives and senior managers to listen to their site operatives and to seek their advice. It empowers site operatives to use their very considerable knowledge and experience to find better and more collaborative ways of enabling the design and construction supply chain team to work together to solve the causes of disruption and reworking.

Performance measurement is the only effective way of exposing the true magnitude and extent of disruption and reworking on site and of exposing the real damage it causes to the morale of the operatives and the friction it generates between the various members of the design and construction team. This incontrovertible evidence gives every firm in the design and construction supply chain a clear, unambiguous and common goal to aim for in their improvement process. All can see the magnitude of the unnecessary costs caused by disruption and reworking and all can see the direct financial benefit that would accrue to their firm if they worked collaboratively to solve the problems that caused the disruption and reworking. Performance measurement also gives them a tool by which they can objectively validate their rate of improvement.

Because performance measurement exposes the magnitude and the extent of disruption and reworking on site, along with the damage it causes to the morale of the operatives and the friction it generates between the various members of the design and construction team, it provides ample motivation to persuade reluctant Chief Executives to divert scarce money and resources into the elimination of the ineffective utilisation of labour and materials.

Performance measurement puts an end to the industry's constant delusions about its true levels of effectiveness. It makes it impossible for senior managers and Chief Executives to claim their firm is achieving labour effectiveness levels of 85% or more when it is only achieving effectiveness levels of 20–30%. Because performance measurement exposes the magnitude and the extent of disruption and reworking on site and targets the improvement process on the causes, it ensures that the design and construction supply chain delivers vastly better whole-life value to the end-user client.

In fact, performance measurement is the only way that vastly better whole-life value can be delivered to the end-user client by the design and construction team.

Finally, performance measurement is the only way by which those who insist their firm is achieving near maximum effectiveness levels can provide the evidence to prove the reality of their claim to others, such as end-user clients seeking better value. No matter how fervently the firm believes performance measurement to be a total waste of time and money because their performance in the effective utilisation of labour and materials is already excellent, the only way they can prove their excellence to prospective end-user clients is to use performance measurement to provide the hard, objective and independent evidence. As the man from the DTI said *'If you don't know how well you are doing, how do you know you are doing well?'*

Since the elimination of unnecessary costs ought to be the primary goal of improvement processes, and since perform-

ance measurement is the only way that the causes of unnecessary costs can be exposed, the following chapters of
this book are essential reading for everyone in every firm in
the construction industry.

A key player from one of the Building Down Barriers pilot
projects recently reminded me of the reason why I came to
believe performance measurement is so important to an
end-user client. Apparently his pilot project team had submitted their detailed proposal and method statements for
their pilot project for the third time and had yet again
explained at length what their firm had done over recent
years to improve its performance and how this would impact
on the pilot project. At this point my normally endless
patience ran out and I apparently lost my temper in a big
way and stormed at them **'Don't keep telling me what
you have f***ing done, show me the f***ing evidence
to prove you've done it'**.

It was then that the pilot project team realised that there
was no evidence available to back up their claims because
none of the firms in the supply-side design and construction
team had ever measured their performance. All they could
offer the end-user client was their unsupported, subjective
assumptions that their improvement programme had delivered better performance.

Case history – M4I presentation

When I was reminded of my unfortunate outburst I immediately recalled a similar experience when I was a Movement for
Innovation (M4I) Board member.

I had attended a meeting in the north-west of England at
which M4I demonstration projects were presenting the improvements they had achieved through the application of
Egan principles (those described in the first Egan report *Rethinking Construction*). Also in attendance with me was a
change expert from the manufacturing industry and towards
the end of the presentations he nudged me in the ribs and
asked me what was missing from the presentations.

When I queried what he meant he said that none of the presentations had mentioned numbers when describing their demonstration project improvements. As a consequence it was his belief that they were all lying about their improvements. He said if a firm truly believed in, and truly understood, the commercial benefit of improvement, it would have measured its performance before and after the claimed improvement and would proudly have announced the resulting figures.

Since none of the firms involved in the demonstration projects had backed up their claims of improved performance with the before and after figures, he was convinced that the claimed improvements were fictitious. In short, because they were unwilling or unable to show us figures to prove the degree of improvement we should ignore their claims and assume they were merely 'talking the talk' and using the buzzwords they thought we wanted to hear, not 'walking the talk' and presenting us with hard evidence to back up their claims.

Finally, we ought to bear in mind the message from Brian Wilson MP, the UK Minister for Construction, in his Foreword to the *Accelerating Change* report:

'Clients need a construction industry that is efficient. An industry that works in a 'joined up' manner, where integrated teams move from project to project, learning as they go, driving out waste and embracing a culture of continuous improvement.'

2 The Unchanged Customer Demand for Improvement

There is a tendency for those involved with the UK construction industry to believe that the demand for radical improvement, that has grown apace since the publication of the Latham report *Constructing the Team* in 1994, is a new phenomenon. In reality, the picture is entirely the opposite with the current demand for radical improvement merely being the latest manifestation of continuous end-user dissatisfaction that can be tracked back at least 70 years.

The only real difference between the current demand for radical improvement from end users and the previous demands is that the current demand is proving far more difficult to misinterpret, ignore or shrug off. The creation by government of a powerful Rethinking Construction organisation to actively champion industry reform, the formation of an active and enthusiastic pan-industry body such as the Design Build Foundation that includes major clients (which then amalgamated with the highly successful Reading Construction Forum to become Be), the use of the Office of Government Commerce (OGC) to champion reform of public sector procurement and the involvement of the external public sector audit bodies (the National Audit Office and the Audit Commission) to force the pace and direction of public sector procurement reform, have more than counter-balanced any attempt by the industry to ignore the end-user demands for radical reform.

Nevertheless, it is important to see the post-*Constructing the Team* drive for radical improvement in its full historical context in order to understand why the current demand for radical reform from end users is not just a short-term aberration. It is also important to understand the key messages that continue to be repeated in every report that has ever been published about the performance of the UK construction industry.

The first major report reviewing the performance of the UK construction industry was produced in 1929 and there have been around 13 similar reports produced between 1929 and 1994. All were inspired by client concerns about the impact of the inefficiency and waste in the construction industry on their commercial performance, and all contained remarkably similar messages.

These client concerns were very effectively summed up in a book written by an architect called Alfred Bossom in 1934 entitled *Reaching for the Skies*. He went to America in the early part of the twentieth century and became closely involved in the design and construction of skyscrapers. This taught him that construction could be treated as an engineering process in which everything is scheduled in advance and all work is carried out to an agreed timetable. The result of using these engineering techniques meant that buildings were erected more quickly than they were in Great Britain, yet cost no more. They yielded larger profits for both the building owner and the contractor and enabled the operatives to be paid from three to five times the wages they received in Great Britain.

On his return, he saw the weaknesses in the performance of the British construction industry with unblinkered eyes and became an enthusiastic advocate for radical change. In his book he stated:

'All rents and costs of production throughout Great Britain are higher than they should be because houses and factories cost too much and take too long to build. For the same reason the building industry languishes, employment in it is needlessly precarious and some of our greatest national needs, like the clearing away of the slums and the reconditioning of our factories, are rendered almost prohibitive on the score of expense.'

'The process of construction, instead of being an orderly and consecutive advance down the line, is all too apt to become a scramble and a muddle.'

'Bad layouts add at least 15% to the production of the cotton industry. Of how many of our steel plants and woollen mills, and even our relatively up-to-date motor works might not the same be said? The battle of trade may easily be lost before it has fairly been opened – in the architect's designing room.'

This description of a fragmented, inefficient and adversarial industry in 1934, which damaged the commercial effectiveness of its end-user clients by being guilty of passing on unnecessarily high capital costs and poor functionality, seems little different to that described in the Latham report *Constructing the Team* in 1994 or the Egan report *Rethinking Construction* in 1998. In fact, the only thing that appears to be different in the 1994 and 1998 reports is the realisation that the maintenance and running costs are also unnecessarily high.

This long historical continuity of end-user client concern about the poor performance of the construction industry is well documented in a book by Mike Murray and David Langford entitled *Construction Reports 1944–98* (see Further reading). In their book's conclusion they state:

'The reports examined in this text have a number of recurring themes that reflect an industry inflicted with long-term illness. The content of many of the reports are strikingly similar and,

indeed, the contributors in Chapter 5 commented to us that the number of similarities with the 1998 Egan report were striking. What is evident, however, is the change in language that spans the reports. The concepts of supply chain management and lean construction are all too evident in the forerunners to Egan, but without the appropriate buzzwords. This is an important aspect of industry change. As a means to combat an apparently volatile and unpredictable market, construction has become more reliant on advice delivered from management consultants and gurus. To some extent this relies on creating a new industry paradigm, one where management is dominant and the use of a new language indicates a commitment to radical change. Sims has argued that the most famous buzzword of all, partnering, has been hijacked by consultants and corrupted by contractors. Furthermore, many of the new ideas are repackaged common sense. This new language may indeed be the building blocks for the twenty-first century construction industry, but critics would argue that too few within the industry can "walk the talk" and even fewer can "talk the talk".'

The reason why the numerous reports between 1929 and 1994 failed to have any impact on the performance of the construction industry is because the industry continues to be blind to its failings. It is unwilling to reveal the truth to itself by measuring its performance, particularly the impact of fragmentation and adversarial attitudes on the effective utilisation of labour and materials and the lack of effective pre-planning of construction activities that concerned Alfred Bossom in 1934.

This situation is made worse because clients continued to reinforce fragmentation and adversarial attitudes by insisting on using a sequential procurement process. This made it impossible to harness the skills and knowledge of the specialist suppliers into design development because they were not brought into the scene until after the construction contractor was appointed and the design was complete. Consequently, it was impossible for them to inject buildability and 'right first time' or greater standardisation of components into the developing design.

Fortunately, the Latham Report in 1994 proved a major catalyst in persuading clients to actively lead the reform movement, rather than standing to one side and expecting the industry to take the initiative. The reason for this radical change in client attitudes was that the Latham Report, for the first time, put a figure of 30% on the cost of inefficiency and waste in the industry. Across the entire construction industry, this burden of unnecessary cost could amount to as much as £17 billion each year. Within the public sector annual expenditure of around £23 billion, it could amount to as much as £7 billion each year.

For individual repeat clients, the message about the high level of unnecessary cost was a powerful driver for them to take a much more active role in industry reform. The Latham Report led to the formation of powerful client groups whose sole intent was to force the pace and direction of reform. The Construction Round Table had been formed in 1992 after the demise of NEDO (the National Economic Development Council), by a small group of major repeat clients such as BAA, McDonald's, Whitbread, Unilever and Transco. The Latham Report findings re-energised Construction Round Table members and encouraged them to take a more active and overt leadership role in the industry, which culminated in the publication of their *Agenda for Change*. The Construction Clients' Forum was formed in 1994 from a mixture of client umbrella bodies, such as the British Property Federation, and major repeat clients such as Defence Estates. The Government Construction Clients Panel was formed in 1997 to provide a single, collective voice for government procurement agencies and departments. In addition to these client groupings, pan-industry groups with dominant client leadership were also formed. The Reading Construction Forum was incorporated in 1995 and the Design Build Foundation was incorporated in 1997.

In 1998, the Egan Report strongly reinforced the concern of clients in the UK about the high level of inefficiency and waste and equally strongly reinforced the earlier Latham Report message of the need for integration. The Egan Report differed from earlier reports by urging the importation of best practice in supply chain management from other sectors. The report stated:

'We are proposing a radical change in the way we build. We wish to see, within 5 years, the construction industry deliver its products to its customers in the same way as the best customer-led manufacturing and service industries.'

Figure 2.1 illustrates the evidence that supports the belief first voiced in the 1994 Latham Report *Constructing the Team*, that at least 30% of the capital cost of construction is consumed by unnecessary costs.

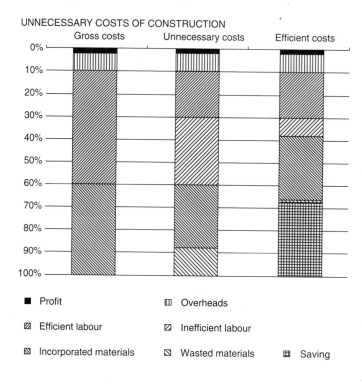

The left-hand column of Fig. 2.1 shows the approximate cost breakdown for a typical construction project involving a fairly traditionally constructed low-rise building with fairly traditional heating and lighting systems. Overheads and profits are around the 10% level, labour accounts for around 50% of the total costs and materials account for around 40%. These figures obviously vary depending on the size and complexity of the building, but in broad terms the breakdown is about right in terms of the proportions of the costs.

The centre column of Fig. 2.1 takes the same overall cost breakdown but separates out the unnecessary costs in the form of the inefficient utilisation of labour and the wastage of materials. The former is caused by things such as reworking, lack of adequate pre-planning, delays with the previous trades, access problems, errors in the drawings, faulty materials, insufficient or inappropriate labour, untidy or cluttered working spaces, delays in the delivery of materials, or changes to the brief. The latter is caused by such things as defective materials, incorrect sizes, scrapped materials from reworking, errors in the drawings, or changes to the brief.

The centre column uses the BSRIA (Building Services Research and Information Association) TN 14/97 *Improving M & E Site Productivity* study (see Further reading) and other similar studies and puts labour efficiency at the industry average of 40% of the labour element of the total costs and materials wastage at 30% of the materials element of the total costs. Obviously these are approximate figures, but the picture they represent is very close to reality and shows very clearly why focusing attention on reducing the unnecessary costs rather than focusing on the profit margin or the overheads would better improve competitiveness. Overall, the ineffective utilisation of labour and the wastage of materials put the total unnecessary cost at around 42%.

The BSRIA study and subsequent work done by the Construction Best Practice Programme and others has demonstrated very clearly that the effective utilisation of labour is invariably around the 30–40% level, with the infrequent examples of best practice rarely exceeding 50%. The point needs to be made that this differs from the perception of senior managers, who believe subjectively that efficiency in the utilisation of labour is far higher and will quote figures as high as 85%. Unfortunately, this perception is rarely supported by accurate on-site measurement and can often be based on a perception as tenuous as 'They must be working efficiently on site because there is no-one in the yard' (this was a comment by a Chief Executive of a specialist supplier firm when asked how he assessed the effective utilisation of labour in his firm).

If the submissions to the various UK construction industry awards are an indicator of the use of measurement to assess effective performance, it is extremely rare for a submission to contain information on performance derived from measuring the effective utilisation of labour or materials. In fact, the concept of unnecessary costs seems to be alien to the construction industry even though it has long been the main concern of firms in other sectors and the reduction of all unnecessary costs has been the primary route by which firms in other sectors improve their competitiveness and their profits.

In the right-hand column of Fig. 2.1, it has been assumed that the improved working practices that come from supply chain integration and a focus on the elimination of unnecessary costs could raise labour efficiency levels to around 70% and reduce materials wastage to around 4%. This gives a saving of around 30%, which is then available to be used for increasing profits, reducing prices, increasing wages, improving research and development and improving training. Obviously these improvements are fairly conservative when compared with the performance of other sectors and if they could be bettered it ought to be possible to achieve savings of around 40–50% as some experts on lean construction have argued.

It is obvious from Fig. 2.1 that the insistence in the Latham Report that improved working practices could save up to 30% of the total cost of construction is well founded.

Despite the intended impact on the UK construction industry of the Latham Report in 1994 and the Egan Report in 1998, there was growing concern by government, by the external auditors of public sector procurement and by a small number of leading edge repeat clients, that the traditional barriers to reform were proving unassailable. It was recognised that the primary reason for this was that the clients (particularly the internal professional advisors within their procurement groups) were refusing to change their traditional, sequential procurement practices and were unable to recognise that this was the main cause of the fragmentation and poor performance of the industry.

This led to three concomitant, but independent moves to re-energise the reform by the publication of three best practice standards for construction procurement that could be imposed on clients by various external means. The three organisations that decided to take this proactive and courageous action were the National Audit Office, the Confederation of Construction Clients and the Cabinet Office (operating through the Department of Culture, Media and Sport, who worked with the Commission for Architecture in the Built Environment and the Treasury).

The Prime Minister became involved with the best practice standard commissioned by the Cabinet Office and he imposed the radical changes in procurement practice (set out in the report *Better Public Buildings* – see Further reading) on the public sector when he launched the report at 10 Downing Street in October 2000. The Confederation of Construction Clients' *Clients' Charter* was launched in December 2000, and independent validation of compliance with the best practice standard set out in the *Charter Handbook* is a condition of attaining the status of a chartered client. The National Audit Office published its report in

January 2001 and it intends to conduct all future audits of central government procurement bodies against the best practice standard set out in *Modernising Construction*. Similarly, *Better Public Buildings* and *Modernising Construction* will heavily influence future audits of local government and health authorities by the Audit Commission.

The importance of the impact of *Modernising Construction* and *Better Public Buildings* on UK public sector clients should not be underestimated. The UK public sector is traditionally responsible for 40% of the total annual expenditure of the construction industry and every public sector client is subject to external audit. As a consequence, the pressure of the external auditors will force a change in the procurement practices of 40% of the industry's clients, who will be required to embrace the best practice standard that is common to *Modernising Construction* and *Better Public Buildings* in order to ensure their procurement practices deliver best value for the tax payer.

These intense external pressures on clients will inevitably ensure that a radical reform of client procurement practice in the UK becomes irresistible. Consequently, there is an urgent need for all involved in construction procurement in the UK to understand the key requirements that are common to all three best practice standards. Whilst this understanding should come from reading the three documents, this chapter endeavours to give busy practitioners in all sectors the priority areas for improvement.

Better Public Buildings

The report's primary thrust is targeted at the functional performance of the completed building and stipulates very firmly that well-designed buildings must enhance the quality of life for the end users. In his Foreword, the Prime Minister stated:

'The best-designed schools encourage children to learn. The best designed hospitals help patients recover their spirits and their health.'

The powerful secondary thrust in the report is the need to achieve better value over the whole lifetime of the building. Again, the Prime Minister stated in his Foreword:

'Integrating design and construction delivers better value for money as well as better buildings, particularly when attention is paid to the full costs of a building over its whole lifetime.'

The report demands of public sector clients radical structural and cultural change in procurement practice, with the most fundamental and far reaching being the requirement (in the 'Why and how' section) that their:

'Procurement arrangements must enable specialist suppliers to contribute to design development from the outset.'

This requirement should come as no surprise to anyone who has read and understood the 1998 Egan Report *Rethinking Construction*, which very clearly saw total integration of design and construction and the use of supply chain management as the key to better value for the end-user client.

The report concludes by listing those actions that must stop and those that must be started. The key actions are:

Stop:
- ☐ regarding good design as an optional extra
- ☐ treating lowest cost as best value
- ☐ valuing initial capital cost as more important than whole-life cost
- ☐ imagining that effectiveness and efficiency are divorced from design

Start:
- ❏ measuring efficiency and waste in construction
- ❏ appointing integrated teams focusing on the whole-life impact and performance of a development
- ❏ encouraging longer-term relationships with integrated project teams as part of long-term programmes, always subject to rigorous performance review
- ❏ using whole-life costing in the value-for-money assessment of buildings

Charter Handbook

The *Charter Handbook* closely follows the theme of *Better Public Buildings* and sets out the obligations that define a best practice client in the Rethinking Construction era. The purpose of the *Charter Handbook* is to set out and describe a best practice standard to which charter clients must commit themselves. This is illustrated in the 'Background to the Charter' section of the *Charter Handbook*, where the Construction Minister, Nick Raynsford MP, is recorded as saying that the charter must set out:

'The minimum standards they (the clients) expect in construction procurement today, their aspirations for the future and a programme of steadily more demanding targets that will drive standards up in the future.'

The *Charter Handbook* recognises that for the current reform of the construction industry to succeed, it is imperative that the clients provide leadership for the essential and radical changes to the structure and culture of the entire supply chain through the reform of their procurement process. The *Charter Handbook* requires chartered clients to lead the drive for continuous improvement of cultural relationships throughout the supply chain and of the constructed outputs of the industry, using performance measurement to provide proof of improvement.

The *Charter Handbook* lists the obligations of a charter client, key of which are the following:

- ❑ Prepare a programme of cultural change with targets for its achievement, over a period of at least three years duration, but preferably five years or more.
- ❑ Measure their own performance in achieving their cultural change programme.
- ❑ Monitor the effects of implementing their programmes of cultural change, by calculating the national key performance indicators that apply to their projects.
- ❑ Review and amend as necessary annually their cultural change programme in the light of what has been achieved.

The *Charter Handbook* requires clients to have procurement processes that deliver (using measurement as the basis of proof) the following key improvements to the constructed products of the industry:

- ❑ major reductions in whole-life costs
- ❑ substantial improvements in functional efficiency
- ❑ a quality environment for end users
- ❑ reduced construction time
- ❑ improved predictability on budgets and time
- ❑ reduced defects on hand-over and during use
- ❑ elimination of inefficiency and waste in the design and construction process

The *Charter Handbook* makes clear that best practice clients should always procure buildings and constructed facilities through integrated design and construction teams, preferably in long-term relationships, which involve all parties in the supply chain in the design process. It also requires the client to enforce the reforms in the structural relationships, culture, process and outputs of the supply-side by making them a condition of any relationship with the construction industry.

Importantly, the *Charter Handbook* also makes clear that consultants (especially architects) must be an intrinsic part of both the industry and the integrated supply chain.

Again, these requirements should come as no surprise to anyone who has read and understood the 1994 Latham report *Constructing the Team* and the 1998 Egan Report *Rethinking Construction*, both of which very clearly saw total integration of design and construction and the use of supply chain management as the key to better value for the end-user client.

Modernising Construction

Although this is the most detailed, comprehensive and specific of the three best practice standards, the *Modernising Construction* theme fully accords with that in *Better Public Buildings* and the *Charter Handbook*. The document is highly critical of the poor performance of the industry and the consequence this has for the public purse. It states:

'In 1999, a benchmarking study of 66 central government departments' construction projects with a total value of £500 million showed that three-quarters of the projects exceeded their budgets by up to 50% and two-thirds had exceeded their original completion date by 63%.'

The document lists and describes the major barriers to improved performance of the construction industry, the key barriers being:

❏ appointing designers separately from the rest of the team
❏ little integration of design teams or of the integration of the design and construction process
❏ insufficient weight given to users' needs and fitness for purpose of the construction

❑ use by client of prescriptive specifications, which stifles innovation and restricts the scope for value for money
❑ design often adds to the inefficiency of the construction process
❑ limited use of value management
❑ resistance to the integration of the supply chain
❑ limited understanding of the true cost of construction components and processes
❑ limited project management skills with a stronger emphasis on crisis management
❑ processes are such that specialist contractors and suppliers cannot contribute their experience and knowledge to designs

The document poses a series of key questions that public sector procurers need to consider if they are serious about improving quality and value for money. The most significant and radical of the key questions are:

❑ Is supply chain integration achieved from the outset of the design process?
❑ Has the whole design and construction team been assembled before the design is well developed?
❑ What are the likely whole-life (running, maintenance and other support) costs?
❑ Have appropriate techniques been used such as value management and value engineering to determine whether the potential for waste and inefficiency has been minimised in the method of construction?
❑ Have efficiency improvements, to be delivered by the construction process, been quantified?

Finally, the document describes those areas where measurement of construction performance is essential, the priority areas to measure being:

❑ the cost effectiveness of the construction process such as labour productivity on site, extent of wasted materials, and the amount of construction work that has to be redone

❑ the quality of the completed construction and whether it is truly fit for the purpose designed

❑ the operation of completed buildings to determine whether the efficiency improvements that the original design was intended to deliver were achieved

Once again, these key questions should come as no surprise to anyone who has read and understood the 1998 Egan Report *Rethinking Construction*, which very clearly saw total integration of design and construction and the use of supply chain management as the key to better value for the end-user client. Similarly, the priority areas where measurement is essential should come as no surprise to anyone who has read and understood the 1994 Latham Report *Constructing the Team,* which focused strongly on the need to integrate the design and construction team in order to eliminate the high levels of inefficiency and waste in the utilisation of labour and materials.

It is obvious from detailed analysis of the three standards that there are two key differences and six primary goals of construction best practice that mark out best practice procurement from all other forms of traditional procurement and which are common to all three standards. These are as follows.

THE TWO KEY DIFFERENTIATORS OF CONSTRUCTION BEST PRACTICE

❑ *Abandonment of lowest capital cost as the value comparator.* This is replaced in the selection process with whole-life cost and functional performance as the value for money comparators. This means industry must predict, deliver and be measured by its ability to deliver maximum durability and functionality (which includes delighted end users).

❏ *Involving specialist contractors and suppliers in design from the outset.* This means abandoning all forms of traditional procurement that delay the appointment of the specialist suppliers (sub-contractors, specialist contractors and manufacturers) until the design is well advanced (most of the buildability problems on site are created in the first 20% of the design process). Traditional forms of sequential appointment are replaced with a requirement to appoint a totally integrated design and construction supply chain from the outset. This is only possible if the appointment of the integrated supply chain is through a single point of contact – precisely as it would be in the purchase of every other product from every other sector.

THE SIX GOALS OF CONSTRUCTION BEST PRACTICE

❏ *'The finished building' will deliver maximum functionality, which includes delighted end users.*
❏ *End users will benefit from the lowest optimum cost of ownership.*
❏ *Inefficiency and waste in the utilisation of labour and materials will be eliminated.*
❏ *Specialist suppliers will be involved in design from the outset to achieve integration and buildability.*
❏ *Design and construction of the building will be achieved through a single point of contact for the most effective co-ordination and clarity of responsibility.*
❏ *Current performance and improvement achievements will be established by measurement.'*

The above six goals of construction best practice were first listed in the UK Construction Best Practice Programme booklet *A Guide to Best Practice Procurement in Construction Procurement* (see Further reading) and were then adopted by the UK Rethinking Construction organisation and promulgated in their publication *Rethinking the Construction Client – Guidelines for construction clients in the public sector'*.

Accelerating Change

In 2002 the Government stepped up the pressure for industry reform by inviting Sir John Egan to examine the performance of the industry four years after the publication of his very demanding report *Rethinking Construction*. The Government also invited him to recommend what measures should be taken to accelerate the pace of reform where he felt acceleration to be necessary. The approach adopted was to set up a Strategic Forum for Construction, chaired by Sir John Egan, to seek hard evidence of real change within the industry. The resulting report *Accelerating Change* was published in the autumn of 2002 and was intended to remind the industry of the reforms first set out in the 1998 report *Rethinking Construction,* to refocus the industry on the areas most in need of improvement and to speed up the lagging pace of reform.

The 1998 report demands that the industry concentrates on the delivery of high quality constructed products for the lowest optimum whole-life costs that fully satisfy the functional and economic needs of the end users over the design life of the constructed product. It insists that this requires integrated design and construction teams drawn from long-term strategic supply-side partnerships that measure and continuously improve their performance to drive out inefficiency and waste in the utilisation of labour and materials. The report makes clear that the long-term strategic supply-side partnerships must be constructed so that the design can be developed with the direct involvement of specialist suppliers and manufacturers. It also makes clear that the logistics of supplying labour and materials to site must be radically improved to drive out inefficiency and waste and achieve a culture of 'right first time'.

In his statement at the start of the report, Sir John Egan concludes by stating:

'By continuously improving its performance through the use of integrated teams, the industry will become more successful. This will in turn enable it to attract and retain the quality people it needs, which will enable it profitably to deliver products and services for its clients.'

The report sets out the key measures that are needed to accelerate the pace of change and the foremost of these is:

'By the end of 2004 20% of construction projects by value should be undertaken by integrated teams and supply chains; and, 20% of client activity by value should embrace the principles of the Clients' Charter. By the end of 2007 both these figures should rise to 50%.'

In its recommendations the report further reinforces its belief in the importance of integration and long-term strategic supply-side partnerships by stating:

'Clients should require the use of integrated teams and long term supply chains and actively participate in their creation.'

The report emphasises the critical importance of accurate performance measurement in any continuous improvement regime by specific reference to it in its vision for the UK construction industry. Within the list of essential actions, the report calls for the following:

□ *a culture of continuous improvement based on perform-*
ance measurement
□ *consistent and continuously improving performance, and*
improved profitability, making it highly valued by its
stakeholders

The report strongly endorses the value of the Rethinking
Construction organisation as the main driver for accelerating the pace of change of the industry and commends its
achievements since it was set up after the first Egan Report.

Rethinking the Construction Client – Guidelines for construction clients in the public sector

Interestingly, just before *Accelerating Change* was published, Rethinking Construction published its guidelines for
public sector clients (which represent 40% of all clients by
value). The published document was entitled *Rethinking the
Construction Client – Guidelines for construction clients
in the public sector* and it broke new ground within the
construction industry by giving a precise and unambiguous
definition of construction best practice. It also used the six
themes of construction best practice that formed the definition to derive six key procurement guidelines for public
sector clients. The Rethinking Construction definition of
construction best practice was taken from the Construction
Best Practice Programme booklet *A Guide to Best Practice
in Construction Procurement* and is as follows:

'*The primary themes of construction best practice are:*

□ **The finished building will deliver maximum function-**
ality and delight the end users.
□ **End users will benefit from the lowest optimum cost of**
ownership.
□ **Inefficiency and waste in the use of labour and mater-**
ials will be eliminated.

❑ *Specialist suppliers will be involved from the outset to ensure integration and buildability.*
❑ *Design and construction will be through a single point of contact.*
❑ *Performance improvement will be targeted and measurement processes put in place.'*

The six guidelines that public sector clients seeking best value should follow when procuring constructed products such as new buildings, refurbishment, adaptations and maintenance are given as:

❑ *'Traditional processes of selection should be radically changed because they do not lead to best value.*
❑ *An integrated team which includes the client should be formed before design and maintained throughout delivery.*
❑ *Contracts should lead to mutual benefit for all parties and be based on a target and whole-life cost approach.*
❑ *Suppliers should be selected by Best Value and not by lowest price: this can be achieved within EC and central government procurement guidelines.*
❑ *Performance measurement should be used to underpin continuous improvement within a collaborative working process.*
❑ *Culture and processes should be changed so that collaborative rather than confrontational working is achieved.'*

Whilst these guidelines are specifically directed at public sector clients, the key measures of the *Accelerating Change* Report makes clear that they apply equally well to private sector clients who are seeking best value in the constructed products they procure. *Rethinking the Construction Client – Guidelines for construction clients in the public sector* makes the very excellent point that:

'*If you don't measure, you will never know how much improvement is possible or desirable. High level targets are the starting point and are necessary to start improvement – but need to be broken down into a series of lower level targets that will enable everyone involved in a project to see how their daily work should be improved to contribute to the overall improvement target. Measures are therefore required in all these areas of activity of the critical processes so that improvement targets can be set and progress to their achievement monitored.*

❑ *The determination to continuously improve overall performance throughout the organisation in a structured way, with everyone wanting to be able to do tomorrow's work better than today's, is also new to many organisations.*

❑ *Ongoing measurement and evaluation of suppliers is essential to maintaining pressure for improvement. Contractors should only retain their framework status if they achieve continuous improvement. On early projects the present levels of performance and the approaches to managing costs collaboratively will be established. On subsequent schemes continuous improvement procedures will be introduced and the resulting performance improvements monitored and documented.'*

In the 1998 Egan report *Rethinking Construction* and in every subsequent publication, accurate performance measurement throughout the supply chain is seen as a critical success factor. Performance measurement provides the evidence to prove how effectively everyone in the supply chain is performing at present and it provides the evidence to prove how their performance is improving on a project-by-project or on a year-by-year basis.

Performance measurement is critical to best practice procurement because it provides hard evidence that the end-user clients can use it to enable them to objectively select the integrated supply-side team that can prove with hard evidence that it has been most effective at driving out inefficiency and waste in the utilisation of labour and materials and can therefore deliver the lowest optimum whole-life cost. This is particularly important because it creates a level

playing field for those firms that have made the heavy investment that is essential to embrace all that is entailed in the
introduction of supply chain management and lean construction techniques within a fully integrated design and
construction team.

The evidence provided by performance measurement enables end-user clients to separate those firms that are merely
talking the talk and using the appropriate buzzwords, from
those firms that are walking the talk and are using supply-
side partnering and supply chain management techniques to
drive out unnecessary costs and drive up whole-life quality
and performance to give end users far better whole-life
value.

Demand-side clients should always bear in mind that performance measurement is the only way that the supply-side
design and construction team can back up any claims they
make about being able to offer an excellent service – *If they
don't know how well they are doing, how do they
know they are doing well?*

3 The Link Between Profits, Competitiveness and Measurement

Any drive for improvement in any activity must start by knowing with absolute certainty where you are with your current performance and where you wish to arrive with your improved performance. As has been pointed out before in this book, unless you know how well you are doing in terms of your current performance, you cannot claim to be doing well because you lack any evidence to prove your claim.

In any business sector, unless you know precisely how good your current performance is (and that of your suppliers) it is impossible to set sensible targets for improvement. If you are close to best-in-class, you only need to improve sufficiently to match the best-in-class performance. If you are well behind the best-in-class, you need to set tough targets for improvement over a short time span. This highlights the importance of benchmarking, or comparing, your performance with that of others in your area of business, especially those that are recognised as the best performers.

However, for benchmarking to be effective you need to be sure that you are comparing like for like so that you are comparing apples with apples and not apples with oranges. The case history below illustrates the massive gap that can exist between reality and assumption and shows just how risky it can be to base an improvement programme on assumptions about your firm's performance, or your suppliers' claimed performance.

Case history – Construction Best Practice Programme exercise

In the autumn of 2002 a *Contract Journal* article described a CBPP exercise that had taken a cross-section of construction industry firms (construction contractors and specialist suppliers) and had compared their subjective views of their effective utilisation of labour with the figures that came from actually measuring what happened on site.

In every case, the firm insisted that it was regularly achieving labour effectiveness levels of at least 85%.

Unfortunately, when reality was carefully measured the picture was far less rosy. Not one of them exceeded 40%, with all but one coming out at between 30% and 40% in their effective utilisation of labour. The exception was measured down as low as 20%, which means the firm in question believed itself to be four times as effective in its use of labour than it really was.

If we also (not unreasonably) assume the poorest performing firm is achieving a wastage of materials due to reworking and other factors that are equally bad and is in the industry worst case band of 30% wastage, their total unnecessary cost is likely to be just over 50%. In other words, their customers are likely to be regularly paying double what they ought to be paying if the firm had achieved the effectiveness levels it claimed.

The Construction Best Practice Programme has endeavoured to help this benchmarking process by publishing a range of Key Performance Indicators and both simplifying and encouraging their use by also making available radar charts and guidance material that enables industry firms to compare their performance with industry norms. In the case of all construction activities, the CBPP offers 10 Key

Performance Indicators that relate to the following areas of performance:

- **Client satisfaction – product.** How satisfied the client was with the finished product/facility, using a 1–10 scale where 10 equals totally satisfied and 1 equals totally unsatisfied.
- **Client satisfaction – service.** How satisfied the client was with the service of the consultants and main contractor, using a 1–10 scale where 10 equals totally satisfied and 1 equals totally unsatisfied.
- **Defects.** Condition of the facility with respect to defects at the time of handover, using a 1–10 scale where 10 equals totally defect free, 8 equals some defects with no significant impact on the client, 5/6 equals some defects with some impact on the client, 3 equals major defects with major impact on the client and 1 equals totally defective.
- **Predictability – cost.** There are two indicators – one for design cost and one for construction cost. Design cost is the actual cost at the available for use stage, less the estimated cost at commit to invest stage expressed as a percentage of the estimated cost at commit to invest stage. Construction cost is the actual cost at available for use stage, less the estimated cost at commit to construct stage expressed as a percentage of the estimated cost at commit to construct stage.
- **Predictability – time.** There are two indicators – one for the design phase and one for the construction phase. Design time is the actual duration at the commit to construct stage, less the estimated duration at commit to invest stage expressed as a percentage of the estimated duration at commit to invest stage. Construction time is the actual duration at available for use stage, less the estimated duration at commit to construct stage expressed as a percentage of the estimated duration at commit to construct stage.

❑ **Profitability.** Profit before tax and interest as a percentage of sales.
❑ **Productivity.** Company value added per employee.
❑ **Safety.** Reportable accidents per 100 000 employed per year.
❑ **Construction cost.** The normalised construction cost of a project in the current year less the normalised cost of a similar project one year earlier, expressed as a percentage of the normalised construction cost of the project one year earlier.
❑ **Construction time.** The normalised time to construct a project in the current year less the normalised time to construct a similar project one year earlier, expressed as a percentage of the normalised time to construct the project one year earlier.

Whilst the Construction Best Practice Programme Key Performance Indicators are a good starting point, it is imperative to ensure the specific goals you set for your improvement process are those that will deliver a constructed product to the end-user customer that precisely matches the actual expectations (not your perceived expectations) of the end-user customer. It cannot be emphasised strongly enough that it is absolutely essential that your goals relate precisely to the true functional and commercial expectations of the end-user customers and not to assumptions made by others on behalf of the end users (such as the property divisions of clients or external organisations such as industry institutions or federations) that may turn out to be erroneous.

Probably the best guide to the true expectations of end users is the Rethinking Construction leaflet *Rethinking the Construction Client – Guidelines for construction clients in the public sector*. Although this was primarily written for public sector clients, its six primary themes of construction best practice are equally applicable to the private sector and are an excellent guide to the true functional and commercial expectations of all end users who are looking to get best value from the construction industry. Rethinking Construction adopted the six themes from the Construction Best Practice Programme booklet *A Guide to Best Practice in Construction Procurement*, which is available free of charge from the Construction Best Practice Programme.

The questions you must ask yourself are what quality, cost and performance do the end users expect from the built product and how does the quality, cost and performance of your built products compare with that of the best-in-class? Only when you have established this firm baseline by measuring your current performance and establishing the expectations of the end user can you develop your improvement programme, set your improvement targets and devise the Key Performance Indicators to check your effectiveness at meeting the improvement targets you have set. These Key Performance Indicators may include the CBPP Key Performance Indicators I have described above, but they may well need to include additional Key Performance Indicators that are more specific to things like the effective utilisation of labour and materials.

To use the very simple example of a 1500-metre runner, the first necessity is to know how fast he or she can run 1500 meters. The next necessity is to know how fast the opposition can run the same distance, especially the best-in-class runners. Having established the gap in performance, the next necessity is to develop an improvement programme to get from where you are to where you want to be. It also helps if a knowledgeable coach or mentor can be employed to help set a sensible and attainable improvement programme and to help check progress against the programme.

It follows from the above that it is imperative is to establish goals for improvement that are targeted at the weaknesses in the performance of the industry that end-users believe are the cause of their not receiving value for money in the built products of the industry. Unless supply-side firms ensure their goals are the same as the demand-side end user's goals, the two sides will be heading in different directions. The publication of the three best practice standards (*Charter Handbook*, *Modernising Construction* and *Better Public Buildings*) now provides a route that can be used to establish the true expectations of the end users and thus avoid the risks inherent in making assumptions about what the end user expects. Whilst reading the three publications is the ideal way of understanding the expectations of the end user, the previous chapters of this book should simplify that process and should enable firms and organisations within each sector of the construction industry to recognise and prioritise those aspects of their working practices that are in most need of reform.

In the last chapter I reminded you that the Construction Best Practice Programme booklet *A Guide to Best Practice in Construction Procurement* had explained that there are two key differences that mark out best practice procurement from all other forms of traditional procurement. I reminded you that these were:

❏ *Abandonment of lowest capital cost as the value comparator.* This is replaced in the selection process with whole-life cost and functional performance as the value for money comparators. This means industry must predict, deliver and be measured by its ability to deliver maximum durability and functionality (which includes delighted end users).

❏ *Involving specialist contractors and suppliers in design from the outset.* This means abandoning all forms of traditional procurement which delay the ap- pointment of the specialist suppliers (sub-contractors, specialist contractors and manufacturers) until the design is well advanced (most of the buildability problems on site are created in the first 20% of the design process). Traditional forms of sequential appointment are re- placed with a requirement to appoint a totally integrated design and construction supply chain from the outset.

This is only possible if the appointment of the integrated supply chain is through a single point of contact – precisely as it would be in the purchase of every other product from every other sector.

The Construction Best Practice Programme booklet also made clear that there are six goals of construction best practice and the six goals require every firm and every organisation involved in the design and construction supply chain to ensure that:

- ❏ *'The finished building will deliver maximum functionality, which includes delighted end users.*
- ❏ *End users will benefit from the lowest optimum cost of ownership.*
- ❏ *Inefficiency and waste in the utilisation of labour and materials will be eliminated.*
- ❏ *Specialist suppliers will be involved in design from the outset to achieve integration and buildability.*
- ❏ *Design and construction of the building will be achieved through a single point of contact for the most effective co-ordination and clarity of responsibility.*
- ❏ *Current performance and improvement achievements will be established by measurement.'*

The six goals of construction best practice, particularly the last goal, demand the use of performance measurement to both establish a firm's precise current performance and to validate the rate of improvement. This applies particularly to the elimination of unnecessary costs in the effective utilisation of labour and materials.

The previous chapter also demonstrated that end-user customer demand for improved performance from the construction industry has changed very little over the last 70 years. This end-user demand for better value constructed products is primarily aimed at improving the competitiveness of the end-user customer. If the traditionally high level of unnecessary costs within the initial capital cost of construction can be radically reduced, the significantly lower initial capital cost will translate directly into lower overheads for the goods or services that are marketed by the end user. This is because the initial capital cost of construction is invariably funded by means of a loan or mortgage that has to be repaid over a number of years and these repayments are likely to form a significant part of the annual overheads. As a consequence of the reduction in the overheads that flows directly from the lower initial capital cost, the end user's competitive position will be improved because the prices charged for the goods or services can be lower and the profit margin will be higher.

Similarly, if the whole-life performance of the constructed product is radically improved by a reduction in the risk of premature failure of components and materials, by a reduction in the annual maintenance costs over the design life, and by minimising the operating costs, the end user's overheads will be further reduced and the uncertain long-term risks that inevitably come from the unforeseen premature failure of components and materials will not have to be allowed for in the overheads.

In the case of the public sector, the decision to proceed with a construction project will have been supported by a detailed investment appraisal over 25 years, or over the design life of the building or facility. Within the investment appraisal the initial capital cost, the annual maintenance costs and the operational costs will have been included with a reasonable degree of accuracy using historic data. If the supply-side team appointed for a given project are able to use supply chain management techniques and performance measurement to drive out unnecessary costs, and are prepared to pass on the majority of this saving to the public sector client so that the initial capital cost is well below the reference cost derived from historic data, it will have the effect of increasing the likelihood of the project proceeding and it will reduce the amount of money that needs to come from the public purse (i.e. from the taxpayer). Traditionally, the public sector has spent just over £23 billion a year on construction activities and Fig. 2.1 in Chapter 2 showed that the elimination of unnecessary costs had the potential to save the public purse around £7 billion a year.

It is obvious from the above why the elimination of unnecessary costs is equally important to demand-side clients in both the private and public sectors. In the case of the private sector, demand-side clients become more competitive; in the case of the public sector, demand-side clients reduce their call on the public purse funded by taxpayers.

It is therefore obvious from the above why one of the end users' primary demands for improvement continues to be a demand that construction industry supply-side firms introduce performance measurement. This is imperative to provide evidence of current levels of unnecessary cost caused by the inefficient utilisation of labour and materials throughout the supply-side team. It is also essential to provide the evidence that the firms that make up the supply-side design and construction team are being successful in their use of supply chain management techniques to eliminate unnecessary costs.

In the past, attempts to reduce the initial capital cost of constructed products have tended to come from the firms involved in the supply-side construction team cutting their profit margin to the bone. All too frequently construction industry firms are working to average profit margins of less than 5% and it is not unusual for firms to tender at cost or below, with the high-risk hope that subsequent claims will inject a minimal profit margin into the project before the final account is settled. An inevitable consequence of this has been the high rate of bankruptcy within the construction industry, a lack of adequate investment in training, low investment in research and development, an increasingly fragmented industry and a growth in adversarial attitudes.

Chapter 2 provided graphical evidence to show that the initial capital cost of construction concealed a magnitude of unnecessary cost that far outweighed the minimal profit margins. In fact, the available evidence suggests that profit margins are traditionally less than 5% and unnecessary costs are around 30% or more. It is fairly obvious from this that construction firms that are intent on increasing their profit margins need to focus their attention very firmly on locating, measuring and eradicating all the unnecessary costs within their own firm and within the firms that make up their supply chain.

In an industry where adversarial attitudes are deeply ingrained, there is a natural tendency for firms to defensively insist that their performance is beyond reproach in order to avoid any risk of being dropped from demand-side client's tender lists or from the preferred lists of their supply-side customers. Whilst a natural fear of the consequences of admitting to being less than perfect might tempt a firm to insist that they have no unnecessary costs because their utilisation of labour and materials is highly efficient, it needs to be remembered that this is a spurious stance unless it can be supported by careful and objective measurement of performance.

In the case of construction contractors, at least 80% of the resources required to construct the facility come from specialist suppliers (sub-contractors, trades contractors and manufacturers) and unless each of the specialist suppliers in the design and construction team is measuring their efficient utilisation of labour and materials, any claim to be highly efficient from the construction contractor is wholly subjective, likely to be wildly adrift from reality and will almost certainly be highly optimistic.

> **Case history – Industry awards submissions**
> I had voiced concern after a construction industry awards assessment meeting about the consequences of the inefficient utilisation of labour and the fact that not one of the candidates had made mention of any improvement in their effective utilisation of labour and materials, or had made mention of

the benefit of measuring their performance. A Chief Executive from a specialist supplier firm that was also on the judges panel said that the attitude of the candidates was entirely reasonable since efficiency levels were universally high across the industry.

He went on to insist that in the case of his own firm, his operatives were achieving (and had always been achieving) near maximum efficiency levels on all the sites on which they were involved. When I queried how he could be sure of his efficiency levels, his response was that there was never anyone in the yard therefore every one of his operatives must be effectively and efficiently utilised on the various sites all of the time.

We then had a lengthy discussion about the value of accurate and objective measurement of on-site performance and the very real and tangible commercial benefit that would automatically come from using the result of accurate performance measurement to drive out unnecessary costs and thus to drive up profits and drive down prices.

He vehemently insisted that the cost of such performance measurement could never be offset against cost savings because it was self-evident to him, as the Chief Executive, that efficiency levels in his firm could not be significantly improved.

Regardless of the end-user client demand that the effective utilisation of labour and materials must be measured so that industry firms know with absolute certainty what level of performance is actually being achieved, there is a powerful commercial reason why industry firms ought to measure their effective utilisation of labour and materials. This applies to every firm on the supply-side and relates to the performance of the firm itself and to the performance of every firm within its supply chain. The reason is that performance measurement can quickly and dramatically improve the competitive position of the firm in question.

The span between tenders on any construction project tends to be fairly narrow and is unlikely to be more than about 5% of the total cost of construction (the tender price). Even in the case of negotiated single tenders, the demand-side customer is likely to compare the negotiated price with a benchmark price derived from an analysis of historical tender prices. Thus any unusually low tenders will alarm the wary demand-side client in case they are due to a major error in the build-up of costs that produced the tender price, or to the firm deliberately submitting a below-cost tender price with the cynical intention of recovering the missing profit through claims. It therefore follows that any construction firm that tenders a price that is below the industry benchmark or below the normal tender span will only be selected if the low tender price can be supported by hard evidence provided by the measurement of improved utilisation of labour and materials.

If the low tender price can be shown to be based very firmly on the use of performance measurement down through the supply chain to significantly reduce the unnecessary costs caused by the ineffective utilisation of labour and materials, and if the firm can show the demand-side customer the before and after evidence, the firm can be reasonably certain of being awarded the contract. If the firm can also show that the use of performance measurement to reduce unnecessary costs has enabled every firm in the supply-side team to significantly enhance their profit margins whilst reducing their prices, the demand-side customer will have the added assurance that there is very little likelihood of the supply-side team indulging in a claims culture to boost the profit margin. Consequently, the demand-side customer can be reasonably certain that the final cost will not exceed the initial cost, as is all too frequently the case where performance measurement is not used to drive out unnecessary costs.

The escalation in price between the tender and the final cost is a major cause of concern for end-user clients, since the escalation is rarely anticipated when the project budget is built into the long-term business plan, or when the detailed investment appraisal is approved and funds are set aside for the project. I mentioned in Chapter 2 the National Audit Office report *Modernising Construction* findings, which said that three-quarters of construction projects exceeded their budgets by up to 50%.

It is obvious from the above that demand-side clients have a strong vested interest in using only supply-side teams that can provide evidence that they are able to offer an initial capital cost that has been driven down solely from the elimination of unnecessary costs, and not by the use of inferior materials or by forcing the specialist suppliers (sub-contractors) to cut their profit margin to the bone. Such hard evidence can only come from the objective measurement of the effective utilisation of labour and materials and a wise demand-side client ought to insist on seeing the measurements from a cross-section of projects over the previous three years, so that the client can be sure that the improvement in performance is being maintained.

A wise client should also demand the names of the firms in the supply-side team for which the measurements are offered, so that it can be seen that the continuous improvement in performance applies to *every* member of the supply-side team that has been put forward for a given project.

It follows that the use of objective performance measurement to improve the effective utilisation of labour and materials in each firm that makes up the supply-side team will vastly improve each firm's ability to compete successfully, no matter what selection system the demand-side customer uses. At the same time, it will also vastly improve the profit margins of each supply-side firm and will vastly reduce the risks of cost escalation that come from the disruption of site activities, such as reworking due to errors on the drawings or defective workmanship, late deliveries of materials, congested or cluttered site conditions, poor programming of site activities, etc.

Obviously, a marked improvement in the success rate in tendering and in profit margins will have the knock-on effect of radically increasing the firm's market share and of increasing share prices.

The previous chapter emphasised the common threads that run through the various reports on improving the performance of the construction industry, written since 1994. These threads were nicely summed up by the UK Minister for Construction, Brian Wilson MP, in his Foreword to *Accelerating Change* where he stated:

'Clients need a construction industry that is efficient. An industry that works in a 'joined up' manner, where integrated teams move from project to project, learning as they go, driving out waste and embracing a culture of continuous improvement.'

The common threads emphasise the critical importance enlightened end-users place on total integration of the supply-side team and the application of supply chain management techniques to drive out unnecessary costs and to drive up the whole-life quality and the whole-life performance of constructed products. The threads also emphasise the importance of driving out the unnecessary costs caused by the ineffective utilisation of labour and materials and converting them into higher profit margins for every supply-side firm.

This chapter has shown how performance measurement can be used to drive out unnecessary costs and convert them into higher profits; it has also explained the very direct link between the introduction of performance measurement and a marked improvement in competitiveness.

However, implementing an improvement programme that is based on performance measurement is only possible if all the firms that make up the supply-side chain radically change the way they do business together. They need to move away from operating in one-off project teams that are selected on the basis of the lowest price, and move towards the formation of long-term strategic supply-side partnerships that over-arch individual projects. These long-term partnerships are essential to a robust and effective continuous improvement regime that is able to seriously reduce the heavy burden of unnecessary costs described in Chapter 2.

Long-term strategic supply-side partnerships create stable working relationships that enable an adversarial, closed-book and unco-operative culture to be replaced by a trusting, open-book and mutually supportive culture where all can admit to where things regularly go wrong without fear of retribution. Such a culture encourages everyone to help each other to find ways of working more effectively together so that 'right first time' becomes the norm for every site and every project. Such a culture recognises the value of the operative as much as it recognises the value of the design consultants and creates a working environment where the operatives are both able, and are encouraged to input their knowledge and experience into the design from the outset.

These long-term strategic supply-side partnerships can best be described as a 'virtual firm' since they are not necessarily bound by contract and they most certainly do not involve one firm taking over all other firms in the supply chain. The concept and operation of a 'virtual firm' is explained in depth in Chapter 5 and is also shown to be a key aspect of the radical improvement process in Chapter 8, dealing with performance measurement at strategic level.

4 The Structure of Performance Measurement

Any book explaining the purpose and the operation of performance measurement at both project and at strategic level needs to explain the meaning of the various terms, such as Critical Success Factors (CSFs) or Key Performance Indicators (KPIs). In my own case, the terms were unknown acronyms before I started the Building Down Barriers development project and were still only vaguely understood when Defence Estates made the decision to embrace the EFQM Excellence Model as the best way of measuring its current effectiveness, of monitoring its rate of improvement and of benchmarking progress in its improved performance with other organisations in other sectors.

DEFINITIONS OF TERMS

The definitions in this chapter are drawn from the EFQM Excellence Model with additional explanations of each term drawn from my experiences on the Building Down Barriers project and at Defence Estates. The terms I have defined are those in use around the construction industry within the current drive for improvement, but I have very carefully restricted my definitions and my explanations to accord with the EFQM Excellence Model since the EFQM definitions are universally recognised across all other sectors and it seems wise to use a language that is common across the rest of the business world.

The terms are as follows.

Benchmark

The EFQM definition is:

'A measured, "best-in-class" achievement; a reference or measurement standard for comparison; this performance level is recognised as the standard of excellence for a specific business process.'

In the case of Building Down Barriers, we recognised that effective supply chain management did not exist at all within the construction industry. If we wished to develop a set of process-based supply chain management tools that construction industry firms could use to drive out unnecessary cost and drive up whole-life quality, we obviously needed to look beyond the construction industry for best practice in supply chain management. As we had no idea where to look for best practice, I approached the Warwick Manufacturing Group for advice and help because they had an international reputation for expertise in supply chain management. I asked them if they could point us to a best-in-class manufacturer that we could use as a benchmark for best practice in supply chain management. We then appointed the Warwick Manufacturing Group to work with us to convert the best-in-class manufacturer's supply chain management process into a set of tools that would enable construction industry firms to import that benchmark supply chain management process.

In the case of Defence Estates, we used the EFQM to locate examples of best practice in the different aspects of our business. In my own case, I discovered that TNT had radically changed the way it used its operations manuals because its EFQM Excellence Model self-assessments had exposed how ineffective and inappropriate the manuals were at the sharp end of their business. Subsequent to the change, the manuals were replaced with simple flow charts that were hung on the wall adjacent to every operation and only the training staff had access to the manuals, which were used to teach the operatives how to understand and use the flow charts. The effectiveness of this improvement had been validated in the subsequent EFQM self-assessments.

Since my area of responsibility at Defence Estates included the production of guides and manuals, I immediately recognised the sense of what TNT had done and also recognised that the lesson was equally applicable to Defence Estates' field of operation, i.e. TNT provided me with an excellent best-in-class standard that I could import into Defence Estates in order to make our guides and manuals more effective at the sharp end.

Benchmarking

The EFQM definition is:

'A systematic and continuous measurement process; a process of continuously comparing and measuring an organisation's business processes against business leaders anywhere in the world to gain information that will help the organisation take action to improve its performance.'

In the case of Defence Estates, we needed to compare our rate of improvement with other organisations, both within the public sector and elsewhere, to see if we were improving as quickly as others, especially other central government departments. The EFQM Excellence Model provides a unified structure to self-assessment and a unified marking system that ensures that the overall self-assessment mark from all kinds of businesses can be compared. Thus we were able to compare our rate of improvement with others and we could be sure that trying to compare the results coming out of totally different measurement systems was not misleading us.

Critical Success Factors

The EFQM definition is:

'The prior conditions that must be fulfilled in order that an intended strategic goal can be achieved.'

In the case of Building Down Barriers, if we wanted to be sure that our supply chain management toolset really worked we had to test and refine the tools on live projects. Consequently, one of our Critical Success Factors (or one of the things that we had to have in position, or have available, before we could achieve our goal of developing a viable and proven supply chain management toolset) was to persuade the army to give us two similar building projects and to guarantee that both would proceed at the same time so that they could be used as test-bed pilot projects for the developing toolset.

Another CSF was the need to be certain that everyone involved, including the two pilot project design and construction teams, had (and continued to have) a common understanding of what the Building Down Barriers project was about so that we were all heading in the same direction and at the same speed. This CSF was checked at six-monthly intervals by holding workshops for the entire team at which The Tavistock Institute and Warwick Manufacturing Group carefully assessed everyone's level of understanding and then ensured any gaps were plugged before the workshop closed.

Yet another CSF was the rate of development of the supply chain management tools. Until The Tavistock Institute and Warwick Manufacturing Group development team had produced a draft tool, the two pilot projects could not move forward. In fact, they could not even start producing the two project briefs until the tool that related to the use of value management was available, because we wanted to ensure that the end users and the specialist suppliers were fully involved in a specific way. Interestingly, the problems the two pilot project teams had in understanding and applying the draft supply chain management tools generated an additional CSF, because the development team had to urgently refine the tools in close collaboration with the two pilot project teams in order to avoid design development on the pilot projects coming to a halt. Consequently, the availability of the refined tools became an additional CSF.

In the case of Defence Estates, the Chief Executive and the senior managers held a series of workshops that were facilitated by an external expert and at which we hammered out the things we had to have available at each stage of the improvement process if we were going to be able to achieve our goals. One of these was the development of a networked user-friendly business management system (BMS) that was loaded with all our new business processes set out as very simple flow charts, each of which could be located and recovered within seconds. We recognised that our improved way of working would not happen until the BMS was in position and could provide our widely dispersed staff with the information they needed to enable them to start changing their outdated working practices to the new working practices, and to do so in a uniform way across the entire organisation.

Key Performance Results

The EFQM definition is:

'Those results that it is imperative for the organisation to achieve.'

In the case of Building Down Barriers, the Key Performance Results related entirely to the deliverables on the two pilot projects. Both had to be completed, be free of defects and ready for operation by the target deadline set by the end users. Both had to provide the end users with an accurate prediction of the cost of ownership over the 35-year life of the buildings, and the predicted cost of ownership (whole-life cost) had to be significantly below the end user's estimated cost of ownership when both were converted into Net Present Value figures (the amount of capital that would have to be set aside and invested to cover the cost of ownership over the 35-year life of the buildings). Both had to deliver all the functional requirements listed and described in the project briefs and do so in a way that ensured maximum functional efficiency, and this had to be confirmed by the end users once they had taken occupation and started using the buildings.

In the case of Defence Estates, the first Key Performance Result was the awarding of the first prime contract. Until a totally integrated design, construction and maintenance team had been awarded the first prime contract, we could not claim to have achieved what we set out to achieve.

Lagging Indicators

The EFQM definition is:

'Lagging indicators show the final outcome of an action, usually well after it has been completed. Profitability is a lagging indicator of sales and expenses. Perception measures are also referred to as lagging (trailing/following) indicators. A perception result relates to direct feedback from a stakeholder, e.g. when employees respond via an internal attitude survey.'

By this definition, most of the Construction Best Practice Programme Key Performance Indicators are clearly lagging indicators. It could be argued that they would be far more effective in driving forward the kind of improvements in performance demanded by end users if the Construction Best Practice Programme Key Performance Indicators had all been the leading indicators I have suggested below instead of the lagging indicators. In the case of Building Down Barriers and its two pilot projects, there were no lagging indicators used because we needed to know what was happening at the time it happened, not what happened a year ago when it was far too late to take action to change and improve the various tool development or design and construction activities. We recognised that it was imperative that we had very rapid feedback from the pilot projects to the toolset development team so that the development team could quickly see what refinements were needed to existing tools, or what additional tools were needed to keep the whole exercise on track.

It seems to me that in any drive to improve performance, lagging indicators are of very dubious benefit because the information they provide always comes far too late to allow those at the sharp end of the improvement process to modify what they are doing where things are not improving as intended.

Leading Indicators

The EFQM definition is:

'Leading indicators, sometimes referred to as driving indicators, are usually measured more frequently than lagging indicators. They are the result of a measurement process that is driven by the organisation itself and is entirely within their span of control, e.g. measuring process cycle times. Leading indicators are those that predict, with a degree of confidence, a future outcome. Employee satisfaction, although a lagging indicator for the morale of the staff, is usually recognised as a leading indicator of customer satisfaction.'

It follows from this definition that leading indicators in the construction industry's improvement drive ought to be those things that should be continuously improved at project level in order to deliver the better value demanded by end users. It therefore makes sense for UK construction industry firms to use as its leading indicators those improvements listed in the *Charter Handbook* (in the USA it would be sensible to use the National Construction Goals). The *Charter Handbook* improvements were described earlier in this book, but as a reminder they are listed here:

❑ major reductions in whole-life costs
❑ substantial improvements in functional efficiency
❑ a quality environment for end users
❑ reduced construction time
❑ improved predictability on budget and time
❑ reduced defects on handover and during use
❑ elimination of inefficiency and waste in the design and construction process

As the *Charter Handbook* gives the above improvements top priority, it would be sensible for project teams to set themselves leading indicators (I've called them Key Performance Indicators in this book to accord with the approach adopted by the Construction Best Practice Programme) that relate very directly to each of the seven *Charter Handbook* priority improvements in performance. Since the elimination of inefficiency and waste dates back to the 1994 Latham report *Constructing the Team*, it would make eminent sense for project teams to measure the efficient utilisation of labour and materials and set themselves a target of reducing it by X% per project as a leading indicator.

At strategic level, construction industry firms ought to set leading indicators across all their projects that relate to the same seven priority areas for improvement so that their leading indicators at all levels are co-ordinated, i.e. everyone in their firm is going in the same direction. Needless to say, the same leading indicators ought to be used by their suppliers (see the EFQM definition of suppliers under 'Supply Chain' below) or chaos will reign.

Partnerships

The EFQM definition is:

'A working relationship between two or more parties creating added value for the customer. Partners can include suppliers, distributors, joint ventures, and alliances. Note: Suppliers may not always be recognised as formal partners.'

From the construction industry's perspective, this definition is primarily about how the supply-side of a constructed product (such as a building) work together within long-term strategic supply-side partnerships to give all customers of that constructed product added value.

Supply Chain

The EFQM definition is:

'The integrated structure of activities that procure, produce and deliver products and services to customers. The chain can be said to start with the suppliers of your suppliers and ends with the customers of your customer.'

Under this definition a supplier can supply either manufactured (or constructed) products or services and thus everyone on the supply-side is defined by the EFQM Excellence Model as a supplier. This is at variance with the somewhat confused practice in the construction industry, where construction contractors generally restrict the use of the term 'supplier' to mean someone who supplies a manufactured or constructed product. Those firms that supply services are referred to using a variety of terms, such as 'sub-contractor', 'trades contractor', 'specialist contractor' or 'consultant'.

Consequently, when someone from outside the construction industry uses the term 'supplier' and assumes the industry will understand it to mean what the EFQM means by the term, confusion can reign. In my view the construction industry would be well advised to start using the same language as other sectors and thus use the term 'supplier' to mean all those within the design and construction supply chain that produce either products or services.

5 The 'Virtual Firm'

Previous chapters made clear that the primary objectives of the reform of the construction industry first mooted by the 1994 Latham report *Constructing the Team* were to drive out the unnecessary costs generated by the ineffective utilisation of labour and materials and to drive up the whole-life quality and the whole-life performance of constructed products.

The scale of this improvement in performance is such that it cannot be done on a single project, but requires the same supply-side team to work together over a series of projects over several years to continuously reform the design and construction process from the lessons learned on each successive project. In an industry where the majority of clients are small and occasional, and the majority of projects are small in value, the industry cannot base the formation and operation of long-term supply-side teams solely on a major client being able to supply a series of similar projects over a period of years.

What is needed is for the supply-side firms to rethink the way they work together so that they are able to come together in long-term supply-side partnerships irrespective of the clients. This co-operative way of working requires the supply-side firms to base their relationships on long-term strategic partnerships or alliances that are mutually

supportive, trusting and open-book. Construction contractors must move away from massive supplier and sub-contractor databases of 20 000 or more firms. Site agents and
buyers must give up the right to go to the market at will in
order to secure the lowest possible price for suppliers and
sub-contractors.

Virtually all the knowledge of how and why things go
wrong on site time after time after time is locked up within
the specialist suppliers (sub-contractors and manufacturers).
If this knowledge is to be released and used to eliminate the
causes of unnecessary costs, it will necessitate the creation of
a totally different relationship between supply-side firms.
The specialist suppliers are not going to admit to construction contractors how much reworking is regularly done or
how much disruption is regularly suffered unless their relationship with the construction contractor is secured by a
long-term partnership or alliance which requires both sides
to be honest and open about what goes wrong time and time
again on construction sites.

The two key differentiators and the six primary goals of
construction best practice described in previous chapters
also demand these long-term strategic supply-side partnerships or alliances. At the end of Chapter 3, dealing with the
link between profits, competitiveness and measurement, the
supply-side relationship that is created by the formation of
long-term strategic supply-side partnerships or alliances was
described as a 'Virtual Firm'. This means that a group of
firms, which constitute the entire supply chain for the design
and construction of a typical building or constructed facility,
must use long-term strategic supply chain partnerships to
form themselves into a stable, mutually supportive supply-
side alliance that works together and operates as a 'Virtual
Firm'.

The concept of a 'Virtual Firm' came out of the pioneer-
ing work on supply chain management done by the Building
Down Barriers process development project, which was
launched in 1997. This was intended to adapt best practice

in supply chain management from manufacturing industry for use in the construction industry. The output from the project was the *Building Down Barriers Handbook of Supply Chain Management* (see Further reading).

When the Building Down Barriers team had to explain how the long-term strategic supply chain partnerships that are the foundation of the Building Down Barriers approach worked, it seemed logical to describe the long-term relationship between the supply-side firms as a 'virtual' relationship since it did not necessarily require a formal contract or sub-contract, nor did it necessitate takeovers or mergers. Others have since undertaken to develop the concept of a 'Virtual Firm', such as the Design Build Foundation (now known as Be), and those wishing to gain from their development work should contact them.

The *Building Down Barriers Handbook of Supply Chain Management* says of long-term supplier relationships:

'Long-term relationships can drive up quality and drive down both capital and through-life costs for clients. At the same time, they can increase profitability for the supply chain. These long-term relationships are likely to be with only a small number of suppliers in each key supply category, because it is not possible to invest in the kind of relationship required with a large number of organisations.'

Recognising and understanding the high level of commonality between apparently differing building types, when they are broken down into components, materials and processes, should help in the formation of these strategic supply-side partnerships. All too often, the cry is heard that 'every building is unique and different', yet when the building is broken down into components and materials a different picture emerges. Steel frames are remarkably common to offices, hospitals, health centres, warehouses, multiple-occupancy living accommodation, libraries, workshops, factories and hotels. Brickwork and blockwork occurs in every building type, from high-rise tower blocks to housing. Windows are common to every building type, with the only real variation being the type of material. Electrical services are also common to all building types, with minor variations where there is a requirement for specialist components. Even mechanical services have a considerable commonality across all building types.

The reality of the benefits that can come from a greater commonality of components and materials across differing building types was picked up in the UK in the 1998 Egan Report *Rethinking Construction*. The report was highly critical of the UK construction industry's unwillingness to grasp the benefits of greater standardisation of components and materials across differing building types. It stated:

'We see a useful way of dealing efficiently with the complexity of construction, which is to make greater use of standardised components. We call on clients and designers to make much greater use of standardised components and measure the benefits of greater efficiency and quality that standardisation can deliver... Standardisation of process and components need not result in poor aesthetics or monotonous buildings. We have seen that, both in this country and abroad, the best architects are entirely capable of designing attractive buildings that use a high degree of standardisation.'

The Egan Report also cited examples of a lack of standardisation of components in the UK:

❑ *'Toilet pans – there are 150 different types in the UK, but only 6 in the USA.*
❑ *Lift cars – although standard products are available, designers almost invariably wish to customise these.*
❑ *Doors – hundreds of combinations of size, veneer and ironmongery exist.*
❑ *Manhole covers – local authorities have more than 30 different specifications for standard manhole covers.'*

Case history – Building Down Barriers

Evidence of what can happen when the specialist suppliers are linked closely with designers and construction contractors within long-term strategic supply chain relationships was clearly shown on the two buildings used to test the application of the supply chain management tools and techniques.

These pilot projects achieved many outstanding improvements in performance and in outputs that came directly from the involvement of specialist suppliers in design from the outset. Not only were these outstanding improvements at project level, the specialist suppliers could see that if they continued to work together with the designers and construction contractors at strategic level, they could continue to improve their performance on a project-by-project basis.

The steel fabrication firm on one of the pilot projects achieved major savings in the capital cost of the steel frame and a major improvement in their profit margin. In addition, they were convinced they could take 15% off the capital cost of any subsequent steel frame if the design and construction team could stay together. The specialist suppliers on both pilot projects became fully convinced of the commercial benefits that could flow directly from enabling them to work with the consultant designers at a strategic level to eliminate the recurrent causes of disruption and abortive work, so that 'right first time' on site can be achieved every time for every project.

At pilot project level, this involvement of specialist suppliers in design from the outset led to a far greater use of standard components and materials, which was not imposed by the supply chain management tools and techniques or by the architect or engineers, or by the end-user client, but came solely from the direct involvement of specialist suppliers and manufacturers at concept design stage.

Examples of the improvements measured on the two buildings that came directly from this way of working together were a 20% reduction in construction time, wastage in the materials due to rework consistently below 2%, labour efficiency (time spent overall on adding value to the building) in the region of 65–70%, no reportable accidents, no claims, an absence of commercial or contractual conflict throughout the two supply chains and a high level of morale on site.

The 'Virtual Firm' always works together when dealing with a client whose procurement process embraces the six primary best practice goals, since such a client would want the same single point of contact and the same efficient supply chain management that would be the norm when buying non-construction products. The client's point of contact with the 'Virtual Firm' should be that which makes best sense to the members of the 'Virtual Firm' and may not be the construction contractor, as it would be in traditional procurement.

For other clients, particularly the small and occasional clients, the supply-side firms in the 'Virtual Firm' act as the situation dictates, either operating as single entities or in cluster groupings, but still giving the client the benefit of the improvements to the construction process that they have developed within their long-term strategic supply-side partnerships. By working together in this mutually supportive way, each firm will be able to ensure and improve its profitability and its market share, no matter what procurement approach the client adopts.

The creation of a 'Virtual Firm' on the supply-side of the industry can be driven by a major repeat client that is determined to force the pace and direction of reform in order to achieve better value from construction procurement and thus reduce its impact on their overheads. There are several excellent examples of this in the UK with clients such as Argent, BAA, McDonald's and several other major retailers that have utilised their skill at managing their retail supply chains to manage their design and construction supply chains.

This passive supply-side response to demand-side pressure from major repeat clients for supply chain integration is neither the only, nor the best way of driving forward the radical reforms that supply chain integration demands. This is because of the risk that supply-side firms may only integrate when working for that specific client and may continue to operate in an inefficient, fragmented and adversarial way for all other clients.

The more effective way of introducing supply chain integration and the 'Virtual Firm' is where the initiative is taken by the supply side and it becomes the way they do their business for all their clients. Where this occurs, supply chain integration has a far greater chance of being introduced because the firms in the 'Virtual Firm' are driven by a mutual recognition of the very real commercial benefits that can accrue directly to them all from the elimination of unnecessary costs and the delivery of better quality. This supply-side initiative tends to be led in the UK by major construction contractors who have recognised that the market is becoming more intelligent and discerning and that those firms who act first to radically improve the performance of themselves and their supply chains stand the greatest chance of maintaining or improving their share of the more discerning market.

This concept of a 'Virtual Firm' may be more easily understood if compared to the operation of a major football team such as Manchester United or a major rugby team such as St Helens. The skills required from those selected for a given game will vary depending on the make-up of the opposing team, and the more closely the skills of the home team can match or exceed those of the opposing team, the greater the likelihood of the home team winning. This necessitates the existence of a squad of players that exceeds those needed for a given game, so that the manager is able to choose a different team make-up for an individual game.

A wise and successful manager will carefully analyse the techniques and skills of the players in the specific opposing team to better understand the precise requirements of the specific game. The wise and successful manager will also carefully study the performance track record of the specific opposing team to assess its strengths and its weaknesses. Having thus carefully established the likely skills requirement for the specific game, the manager will compare those precise requirements with the skills, experiences and temperaments of the individual players in the full squad (which may well be double or treble the number of players needed for an individual game) and will pick a team, plus a small number of reserves, for the given game.

From this analogy it can be seen that the operation of the 'Virtual Firm' is closely akin to that of a major football or rugby team. The full squad of strategic supply chain partners will need to be a good deal more than would be required for any given project, because it must encompass the full range of skills and experience necessary for the wide variety of design and construction needs that will occur across the full range of building and civil engineering projects that the 'Virtual Firm' might wish to embrace. From this full squad, the 'Virtual Firm' selects the team that is appropriate for a given project. This will need to include a small team of reserves to cover situations that might arise during the design and construction process, e.g. a switch from steel frame to concrete frame, or an unexpected and unforeseeable overload of an individual firm for reasons outside the given project.

There is another aspect of the management of a successful football or rugby team that has great relevance to the management of a successful 'Virtual Firm'. Consistently high performance of the team would be impossible if the players were constantly changed for every game, with not even the opportunity to practise together in the relatively stable, full squad. There would be no empathy or loyalty between the players, they would have virtually no understanding of each other's skills, experience and temperament, they would never have had the opportunity to regularly train or play together as a co-ordinated and mutually supportive team. The critically important need to develop team skills, loyalty and a common goal for the team (which must apply to the full squad as much as the team for a given game) would be impossible. The manager would find the selection of the most appropriate and effective team for a given game an endless and thankless task and would probably be driven to selecting the team for a given game with a pin (or by use of lowest price tendering from specialist suppliers in the case of the construction industry).

The successful team requires the manager to have a full and detailed knowledge of every aspect of the skills and experience of the full squad. The manager must also have total confidence in the loyalty of every member of the full squad and must have total belief in their shared understanding of mutually agreed goals. With this firm foundation for the selection process, the manager can be reasonably certain that the team selected for any given game can be relied on to perform as an effective, efficient, enthusiastic and mutually supportive team.

These considerations are equally relevant to the effectiveness and success of the 'Virtual Firm'. All too often projects suffer because the design and construction team are cobbled together for the first time and have no expectations of ever being together in the future. Worse still, most of them will have been selected on a lowest price basis, where profit margins have been squeezed to the bone and the only way of making a decent profit may well be through claims against other team members, or against the client.

Even worse is the fact that many team members will be introduced long after the game has started, since most specialist suppliers (sub-contractors, specialist contractors or manufacturers) will not be appointed until the design is well advanced or even complete. Consequently, their skill, experience and knowledge will be ignored in the development of the design, even though harnessing it from the outset of design development would most certainly have improved the cost effectiveness and the buildability of the constructed solution and thus improved the profits of all concerned.

Any hope of creating a mutually supportive, loyal and highly skilled team in this environment is clearly an impossibility, and the resulting fragmentation and adversarial attitudes is the cause of the situation which is very accurately portrayed in the National Audit Office report *Modernising Construction* with the statement:

'In 1999, a benchmarking study of 66 central government departments' construction projects with a total value of £500 million showed that three-quarters of the projects exceeded their budgets by up to 50% and two-thirds had exceeded their original completion date by 63%.'

If all of us can understand and appreciate what makes a successful team in top level football or rugby, why can we not transfer that understanding to our own construction industry? In the case of the construction industry, 80% of the team members are drawn from the specialist suppliers sector of the industry, and because they are not part of a 'Virtual Firm', they are generally selected on a project-by-project basis by the lowest price they can tender for the individual project. Consequently, their long-term security and profitability are high risk, the rate of bankruptcy is far higher than in other industries, the entry level is dangerously low and the valuable skill and experience of the specialist suppliers is rarely ever harnessed to drive out unnecessary costs and drive up quality.

How often do we as individuals join the endless and constant chorus of complaints about 'cowboy builders' or 'cowboy suppliers' in our domestic lives, but operate in a manner that enables the existence of such firms when we switch to our corporate lives?

We all need to rethink the design and construction process and replicate the experience of successful firms in other sectors. They have demonstrated quite clearly that their success is founded on long-term strategic supply chain partnerships that embody the seven principles of supply chain management described in the Building Down Barriers *Handbook of Supply Chain Management*. The 1998 Egan Report *Rethinking Construction* made very clear that integration of design and construction and the use of best practice in supply chain management must also be the foundation of successful 'virtual firms' in the construction industry. The Egan Report stated:

'We are proposing a radical change in the way we build. We wish to see, within 5 years, the construction industry deliver its products to its customers in the same way as the best customer-led manufacturing and service industries.'

The seven universal principles of supply chain management are described in depth in the Building Down Barriers *Handbook of Supply Chain Management*. The handbook explains that there is one primary, over-arching principle of effective supply chain management and six supporting principles. These are briefly as follows.

Primary principle

❑ **Compete through superior underlying value.** Key members of the supply-side design and construction supply chain work together to improve quality and durability, and to reduce underlying unnecessary costs (the labour and materials elements of the component and process costs) while improving profits. The reduction in unnecessary costs is primarily about ending disruption and reworking on site and the achievement of a 'right first time' culture throughout the supply-side design and construction team. Key to this principle is the close collaboration between the design professionals and the specialist suppliers (sub-contractors, trades contractors and manufacturers) that can only come through long-term strategic supply-side partnerships.

Supporting principles

❑ **Define client values.** This requires all members of the supply chain (from end users to manufacturers) to work together, using formal value management techniques, to define and record the detailed business needs of the end users that must be delivered efficiently by the built solution. This ensures that the specialist supplier's operatives who are constructing the built solution have a detailed understanding of the end user's functional requirements.

❑ **Establish supplier relationships.** The products and services of the specialist suppliers (sub-contractors, trades contractors and manufacturers) account for over 80% of the total cost of construction. It is therefore essential for the entire design and construction supply chain to establish better and more collaborative ways of working, so that the skills throughout the supply chain can be harnessed and integrated to minimise waste of labour and materials. These better and more collaborative ways of working should also encourage exploitation of the latest innovations in equipment, materials and building processes.

❑ **Integrate project activities.** This involves breaking down the construction activities into sub-systems or clusters. These are relatively independent elements of the whole building or facility, such as groundworks, frame and envelope, mechanical and electrical services or internal finishes. Within each sub-system or cluster, the design, construction, material and component suppliers work together in close collaboration to develop detailed designs, construction methods and actual prices for delivery.

❑ **Manage costs collaboratively.** This involves 'target costing' where suppliers work backwards from the client's functional requirements and gross maximum price (maximum affordable budget). The supply chain, particularly in the cluster groupings, work together to design a product that matches the required level of quality and functionality and provides a viable level of profit for all at the agreed target price (which must be within the gross maximum price).

❑ **Develop continuous improvement.** The specialist supplier members of the design and construction supply chain (sub-contractors, trades contractors and manufacturers) openly measure and assess all aspects of their current performance, especially their effective utilisation of labour and materials. The entire design and construction supply chain then agree continuous improvement targets for each firm's design or construction performance that will deliver maximum savings in underlying process and materials costs. The ultimate goal is to eliminate all the unnecessary costs that are caused by the ineffective utilisation of labour and materials.

❑ **Mobilise and develop people.** All involved must recognise that their staff will need to learn new ways of thinking, acting and reacting. This involves unlearning old ways and recognising the challenges to be met and the resistance and difficulties that can be expected.

At its simplest, strategic supply chain partnering (or lean construction) is the means by which the supply-side firms work together to drive out all forms of unnecessary cost and to drive up the quality of the constructed product. It is the foundation of every supply-side firm's ability to compete effectively for work in any situation.

When we buy a manufactured product from any other sector (such as a television, a car or a ship) we do not expect to have to enter into a partnership arrangement in order to ensure value for money. In most cases (as in the construction industry, where the majority of clients are small and occasional) the purchase will be one-off and any form of partnership between the client and the supply-side would be of very limited value in driving out unnecessary costs.

There will, of course, be the occasion where a large number of identical or similar products will be required over a period of time by an individual client and this may well make a partnering arrangement sensible for a particular client in a particular instance. Nevertheless, the lesson from other sectors and the message from the 1998 Egan Report *Rethinking Construction*, is that partnering will deliver the greatest improvements in performance where it is the basis of the long-term strategic relationships between firms on the design and construction supply-side of the industry.

This necessitates a radical and profound change in the way the supply-side firms operate and this in turn requires their Chief Executives to understand the nature of the changes in working practices that must be put in place within their own firm and within the firms with which they do business. The Chief Executives must also measure their organisation's current performance (such as the effective utilisation of labour and materials) and that of their suppliers, so that they have a firm basis from which to start the improvement process. They must then become fervent champions for the changes in working practices, because only powerful and clear-sighted leadership from the Chief Executive can make those changes happen.

The magnitude of these changes should not be underestimated; they will affect everyone in the design and construction supply chain and they will not happen without a major change programme and the investment in carefully structured training and mentoring. Measurement of performance will be difficult at first since it has rarely been done in the construction industry and the results may be hard to accept, both for the organisation and for the individual concerned. This will be especially so where it relates to the effective utilisation of labour and materials, and the initial measurements validate the low levels assumed by the Latham Report *Constructing the Team* and confirmed by the work of the Building Research Establishment CALIBRE team and by Building Services Research and Information Association Technical Note TN 14/97.

Many people will find the changes threatening and will endeavour to thwart them and maintain the status quo. Because of the heavy baggage many people carry of established and comfortable custom and practice, they will find it difficult to understand the reason for the new ways of working and this will require the Chief Executive to ensure that the message is expressed in simple, easy-to-understand terms and is constantly reinforced.

The lesson from those organisations that have successfully and radically improved their performance is absolutely clear. Any drive to radically improve performance will not be successful unless it is led by a Chief Executive who obviously understands the nature and magnitude of the changes in working practices and can be seen to be clearly and implacably determined to make those changes happen. This lesson cannot be overstated.

Radical change of the magnitude needed by construction industry firms that are intent on embracing construction best practice, as is now defined by Rethinking Construction's six themes of construction best practice, will be impossible unless the radical changes in working practice are very overtly 'owned' by the Chief Executive in person.

This will pose the greatest burden squarely on the shoulders of the Chief Executive. Without effective and determined top-level leadership, without a shared understanding between the Chief Executives of the firms that need to work together within the long-term strategic supply-side partnerships of what needs to be changed and why it needs to be changed, it will be impossible for anyone below the Chief Executive to instigate and enable the radical changes that are necessitated by the six goals of construction best practice.

The critical importance of the Chief Executive's role was illustrated very potently by the feedback from a series of regional workshops held by Rethinking Construction in the autumn of 2002. The workshops were entitled 'The National Debate', were aimed at public sector clients and their construction industry suppliers, and their purpose was to assess how well the Rethinking Construction reforms were actually progressing down at grass root level beyond the high level rhetoric. The workshops were also to discover what was creating barriers to progress and preventing clients from radically changing and improving their procurement processes. The same message came out of every workshop and can be summed up as follows:

The pace of reform is seriously hindered, and in many cases halted, because it lacks powerful and committed leadership.

In the following chapter I explain what needs to be done to provide this powerful and committed leadership.

6 Effective Leadership

Previous chapters have described the demand-side led internal and external pressures that have been forcing all sectors of the UK construction industry to embrace the radical and profound change necessitated by construction best practice. These pressures were first inspired by the 1994 Latham Report *Constructing the Team* and have been reinforced by a series of subsequent reports such as *Rethinking Construction*, *Modernising Construction*, *Better Public Buildings*, *Charter Handbook* and *Accelerating Change*.

The earlier chapters described the very real commercial benefits that can accrue from an integrated supply-side that applies the techniques and tools of supply chain management, in terms of higher and more assured profits, better whole-life value and lower prices, and they also explained how the supply-side needed to work together within long-term strategic supply-side partnerships.

However, the last chapter also stated categorically that no matter how intense the external pressures for change may be, radical and profound changes within an individual firm or organisation will be impossible without a Chief Executive who clearly understands the nature and purpose of the changes and very overtly and directly champions them.

This applies whether the firm or organisation is on the client's demand-side or on the industry's supply-side.

The Chief Executive must ensure that everyone in the organisation and everyone in the organisations with which it has business links, understands what working practices are to change, how they are to change, why they are to change, when the changes must be implemented, how the improvements in performance will be measured and what commercial benefits the changes will deliver.

Whilst the change process must be initiated and led by the Chief Executive in person, it is also true that a major change in working practices will not happen without a great deal of concerted effort on the part of everyone in the organisation, and an acceptance by everyone that all will have to change the way they work to some degree, with some having to change profoundly. Radical change can easily be thwarted by the inherent and powerful inertia of established custom and practices, and by covert resistance at all levels, especially at senior management level.

It must be obvious to all that the change process is owned, structured and directed by the Chief Executive in person, whose every action and every spoken and written utterance constantly reinforces the direction and the urgency of the change. The experience of organisations that have successfully embraced radical and profound change teaches that the change process needs to be focused into four key areas if it is to be successful. Without all four being in place and operating concurrently, the change will become nothing more than wishful thinking and will soon be forgotten and replaced by the next bright idea.

The four essential and interlinked ingredients of successful change are as follows:

❑ A clearly explained and rational goal that all can understand, with which all can identify, and which can be related to specific improvements in performance that can be measured and compared with current performance.

❑ Committed, determined and overt leadership by the Chief Executive which leaves no-one in any doubt about where the change must take the organisation, why it is commercially essential to go there, and the timescale for the change.

❑ A detailed and comprehensive action plan for the development and implementation of the changes in working practices which explains in simple, easy-to-understand language what must be done differently by every member of the organisation. Adequate and appropriate training must support this for those who are required to operate in a different manner.

❑ A simple and· easy-to-understand explanation of the commercial benefits that will be delivered by the changes in working practices. This is best expressed in terms which relate to improved product quality, improved efficiency in working practices, reduced waste in the production process (both labour and materials) and, most importantly, reduced costs and increased profits.

In any organisation, in any business sector, major changes in working practices are extremely difficult to initiate and achieve. Evidence shows that it is rare if as many as 30% of the workforce are in favour of the changes where they affect their own working practices. Another 30% will fight against the changes (usually in covert ways which will be concealed from the Chief Executive) because they are afraid that the changes will adversely affect their status, their ability to perform, or their pay. The remainder will sit on the fence until they are convinced the changes are inevitable and beneficial.

Those covertly against the changes in working practices will include many at middle and senior management level, including some at board level. They are generally older and inevitably carry more baggage, in the form of being more wedded to the familiar, comfortable and trusted ways of working. Quite often, they are also not unsurprisingly worried that they are going to find it difficult to learn the new and unfamiliar ways of working at their time of life, and that this is likely to mean that their position in the hierarchy of the organisation will suffer as the younger and more junior staff more quickly learn the new ways of working and see their superiors struggling to cope.

It can be seen from the Aerospace Industry case history in Chapter 8 that two important ingredients of a successful change process are communication and education. It is totally unreasonable to expect people to embrace radical change unless it has been explained in a language they can understand and has been illustrated by examples drawn from their own day-to-day working practices. It is equally unreasonable to expect people to embrace radical change unless they have been given adequate training in the new working practices. It is not acceptable to assume the explanation is adequate without verifying if it has been understood at all levels (with the responsibility for selecting the most appropriate language being that of the sender of the message). It is also imperative to measure whether the training has actually changed working practices, preferably by the use of an objective, bottom-up feedback mechanism that has a good track record of success.

Those tasked with communicating within the change process (including the Chief Executive) should bear in mind that it has always been a reality of buying and selling that you can generally buy using your own language, but you can rarely sell unless you use the language of the potential buyer. Consequently, those 'selling' the message about radical change within an organisation need to use the language of those that need to 'buy' the message and this may require the message to be differently phrased for different recipients.

An excellent tool to measure and test the rate of improvement and to provide well-structured support for the change process is available in the UK in the form of the European Foundation for Quality Management (EFQM) Excellence

Model. This has an outstanding track record of success across the private and the public sector, and in large as well as small organisations in the UK and across Europe. The 'Award Simulation' self-assessment system provides an excellent mechanism for supplying a regular, annual, consistent and very objective measurement of the rate of improvement from the people at the sharp end of the organisation. It tells the Chief Executive and the board what is really happening at the workface, and it provides the workforce with an opportunity of ensuring that the truth about what is going wrong in current working practices will get to the Chief Executive and the board without being filtered or massaged by middle and senior managers.

The EFQM Excellence Model also forces organisations to adopt a very structured approach to improvement by the use of nine inter-dependent criteria:

❑ **Leadership.** How leaders develop and facilitate the achievement of the mission and vision, develop values for long-term success and implement these via appropriate actions and behaviours, and are personally involved in ensuring that the organisation's management system is developed and implemented.

❑ **People.** How the organisation manages, develops and releases the knowledge and full potential of its people at an individual, team-based and organisation-wide level, and plans these activities in order to support its policy and strategy and the effective operation of its processes.

❑ **Policy and strategy.** How the organisation implements its mission and vision via a clear stakeholder focused strategy, supported by relevant policies, plans, objectives, targets and processes.

❑ **Partnerships and resources.** How the organisation plans and manages its external partnerships and internal resources in order to support its policy and strategy and the effective operation of its processes.

❑ **Processes.** How the organisation designs, manages, and improves its processes in order to support its policy and strategy and fully satisfy, and generate increasing value for, its customers and other stakeholders.
❑ **People results.** What the organisation is achieving in relation to its people.
❑ **Customer results.** What the organisation is achieving in relation to its external customers.
❑ **Society results.** What the organisation is achieving in relation to local, national and international society as appropriate.
❑ **Key performance results.** What the organisation is achieving in relation to its planned performance.

Of the nine EFQM Excellence Model criteria, the most important are leadership, processes, customer results and key performance results and this is reflected in the marking system that gives the following share of the total marks available:

❑ **Leadership** 10%
❑ **Processes** 14%
❑ **Customer results** 20%
❑ **Key performance results** 15%

At the start of this chapter four key essentials of any successful change process were listed:

❑ a clearly defined goal for the change process
❑ committed leadership of the change process from the Chief Executive
❑ well-defined and clearly described processes for the improved working practices
❑ a clear explanation of the commercial benefits the change process will deliver

Not surprisingly, these four essential ingredients have a close correlation with the four most important criteria from the EFQM Excellence Model. Leadership is about setting a clear and unambiguous goal for the organisation and communicating that goal to everyone in a language each can understand. Process is about providing everyone with a route map that enables each of them to modify or change their outmoded working practices for best practice in a consistent way across the entire organisation. Customer results and key performance results are about objectively measuring the commercial benefits that are delivered by the changes in working practices.

The importance of powerful and committed leadership cannot be overstated. The overwhelming evidence from those organisations in every sector of private and public activity that have successfully embraced radical change is absolutely clear: no matter how much superficial enthusiasm there is for radical change, it will be stillborn unless the Chief Executive is seen to be totally committed to the change. This commitment must include the Chief Executive giving a very simple and straightforward explanation of the goal for the change process that must be expressed in terms that everyone (without exception) can understand, and there must be some way of testing the understanding at an individual level across the organisation.

Feedback from organisations that have successfully achieved radical improvement warns that making assumptions about what people understood can be very misleading. It also carries a high risk that whilst everyone overtly claims to have understood the message, the hidden reality is that everyone goes off in different directions because their interpretation of the message had been slightly different. It is imperative that their understanding is tested and validated, and the message must be rephrased in more appropriate language if it is discovered that it has been interpreted differently across the organisation.

The change will also be stillborn if it is unwisely assumed that everyone will be able to modify his or her entrenched traditional working practices without considerable and appropriate help and education. The education needs to be carefully crafted and structured to address the areas where the working practices of the organisation do not match the best practice of the three standards (*Modernising Construction, Charter Handbook* and *Better Public Buildings*).

The Construction Industry Training Board (CITB) in the UK is liaising with various organisations to develop appropriate training workshops and courses. A good example of this is the ICOM/CITB link-up that offers a Diploma course in Construction Process Management (see Further reading) that is built around the six goals of construction procurement best practice from the Construction Best Practice Programme booklet *A Guide to Best Practice in Construction Procurement*. The CITB is also working with ICOM to develop layered action learning packages that start with operatives and end with senior managers and are also based on the six goals of construction best practice.

Experience has shown that the inertia of deeply embedded traditional working practices could well be powerful enough to neutralise the change process, no matter how apparently beneficial it appears to the Chief Executive. Consequently, the Chief Executive will need to appoint a team to both develop the new working practices and then to work with the staff and operatives to ensure the application of those new working practices throughout the organisation. To ensure the maximum effectiveness of the team, it will almost certainly need to work directly to the Chief Executive so that everyone in the organisation is left in no doubt about the Chief Executive's intentions and there is no possibility of the team being side-lined by those senior staff that are covertly against change.

As was made clear earlier in the chapter, of paramount importance to the successful adoption of the new working practices will be the provision of appropriate training and coaching in their appreciation and application. It is unreasonable and absurd to assume that people who have always worked in one way can suddenly work in a radically new way without a great deal of help and support. This must take the form of providing them with a detailed route-map for the process that constitutes the new working practice, and training in its use by trainers that fully understand the new processes, the improved performance the new processes are to deliver, the way the improved performance will be measured, the commercial pressure that necessitated the new processes and the impact the changes will have on the services and constructed products they must deliver to the customers (internal and external customers). It follows from this that the first people that need to be trained in the new, improved working practices are the firm's training staff. Unless they are walking the talk there is absolutely no possibility of those they train walking the talk!

In order to force the pace of change and test if everyone is really walking the talk, rather than just using the appropriate buzz words, the Chief Executive must ensure that there is some way of measuring the improvements in performance that ought to be delivered by the new working practices. The result of the regular measurement of improved performance needs to be communicated in simple, straightforward terms to everyone in the organisation at reasonably frequent intervals. Ideally, this should name and shame as well as giving public recognition to those that have been most successful in applying the new working practices and improving their performance.

It has always been said that 'success breeds success' and this is equally true of organisational change. As long as those involved in operating the changed working practices can see real and tangible improvements in the products those new working practices are delivering to their customers (and these can be internal as well as external customers), they will be motivated and enthused to continue to do things in this new way. Consequently there will be far less chance of the inertia of traditional working practices dragging them back to the old and inefficient ways of working.

Case history – Building Down Barriers

Whilst the morale of those at the sharp end will be boosted when they are an intrinsic part of the change process that delivers improved products to the client, the converse is true and their morale can be devastated if the changed working practices are not sustained.

In the case of a steel fabrication firm involved on one of the two Building Down Barriers pilot projects, the morale of the fabricators and erectors was sky high at the completion of the Building Down Barriers project. Their skill and experience had been harnessed from the outset of the design of the steel frame and they had therefore been very much instrumental in achieving 'right first time' in the fabrication and erection of the steel frame.

The erectors had been able to get the architect and the structural engineer to understand the erection problems that were encountered time and time again and they were convinced that their input at the commencement of design development was one of the main reasons why the steel frame had been erected 'right first time' for the first time in their experience. The fabricators had also been able to get the architect and the structural engineer to understand the fabrication problems that were encountered time and time again and they were also convinced that their input at the commencement of design development was another of the main reasons why the steel frame had been erected 'right first time' for the first time in their experience.

Unfortunately, the subsequent project was a conventionally procured warehouse and their skill and experience was not utilised by the consultant designers despite intense lobbying by their Chief Executive and warnings from him that the design was flawed. The outcome was a steel frame where the design was riddled with the usual errors and the fabricators knew that parts of the steel frame they were fabricating would have to be taken down after erection and would have to come back to be modified. The erectors also knew that they would have to dismantle parts of the frame after erection, take them back to the factory for modification, and then return to site to re-erect them.

Needless to say, their Chief Executive said their morale that had been lifted to unprecedented heights by their Building Down Barriers experience, but was devastated by their subsequent warehouse experience. Yet their Chief Executive said the warehouse job was quite normal and the flaws in the design of the steel frame were no worse than usual.

The problem was that the erectors and fabricators had assumed that what happened on the Building Down Barriers project was so sensible and beneficial that it would be replicated on all subsequent projects for all other clients.

It follows from the above that unless the Chief Executive is prepared to be seen by every member of the workforce to be totally committed to the change programme, unless the Chief Executive is prepared to explain the goal for the change programme in a simple and easy to understand language, unless the Chief Executive is prepared to ensure the new tools (working practices) are developed and applied, unless the Chief Executive is prepared to ensure appropriate training is available and that staff are able to attend it, unless the Chief Executive is prepared to ensure the improvements in performance are regularly measured and the results published to the workforce, unless the Chief Executive believes the changes should be the way the organisation operates for all its customers, then there is little point in even thinking about introducing a radical change in the way the organisation operates.

This all-important leadership role of the Chief Executive applies to every firm or organisation in the design and construction supply chain. It applies equally to the end user as it does to the manufacturer, since the application of the two key differentiators and the six primary goals of construction best practice from the three best practice stand-ards require radical changes in the way all of them operate.

The previous chapter explained the operation of the 'Vir-tual Firm' that is created by the long-term strategic supply-side partnerships that are an essential part of supply chain management. It is imperative that every Chief Executive of every firm within the 'Virtual Firm' has a shared understand-ing of how the working practices need to be changed, why they need to be changed and how the improvements will be measured. It is also imperative that they are all equally committed to making the changes happen and that they are willing to be open with each other about what changes are going well in their firm and what is not going so well.

This all-important leadership role also applies to the vast number of institutes and trades organisations within the UK construction industry. They too must become an intrinsic, co-ordinated and supportive part of the change process if the best practice approach of the three standards is to take hold and flourish. Their Chief Executives must provide co-ordinated leadership for the new, integrated way of designing and constructing which delivers the six primary goals of best practice. Unless they do so, the traditional fragmentation and confusion will continue, where each in-stitute and each trades organisation appears to be heading in a different direction with a different view of what constitutes best practice.

This need for leadership from the pan-industry organisa-tions tasked with leading the drive for radical reform was very effectively grasped and demonstrated by Rethinking Construction in the autumn of 2002 when they published guidelines for public sector clients and based their six goals of construction best practice on the six goals of construction procurement best practice from the CBPP booklet *A Guide to Best Practice in Construction Procurement*.

This use by Rethinking Construction of the CBPP six goals to give a very precise universal definition of what constitutes best practice, set an excellent example for the industry as a whole and is a move that the many institutions and federations across the industry ought to replicate. The six goals from the CBPP booklet became the Rethinking Construction *'six primary themes of construction best practice'* and are:

❑ *'The finished building will deliver maximum function-ality and delight the end users.*
❑ *End users will benefit from the lowest optimum cost of ownership.*
❑ *Inefficiency and waste in the use of labour and mater-ials will be eliminated.*
❑ *Specialist suppliers will be involved from the outset to ensure integration and buildability.*
❑ *Design and construction will be through a single point of contact.*
❑ *Performance improvement will be targeted and meas-urement processes put in place.'*

Effective leadership is not an issue constrained to the con-struction industry; it must also include the further education establishments that provide courses in the various aspects of design and construction. Chief Executives or heads of fur-ther education establishments must also play their part in supporting the change process. They must ensure their vision for the culture, structure and working practices of the industry matches that of the three standards (*Modernis-ing Construction, Charter Handbook* and *Better Public Buildings*) and their six goals of construction best practice.

They must ensure that the graduates they produce share a common vision of the reformed industry that accords with the direction those at the leading edge of reform have taken and accords with the true expectations of the end-user clients. They must abandon any assumptions they have that might continue to reinforce the traditional fragmenta-tion and adversarial attitudes of the industry, and they must actively reinforce the direction and the pace of reform towards the goal of total integration of design and construction, the elimination of all unnecessary costs in the design and construction process and the delivery of best whole-life value and best whole-life performance.

Most important, those at the head of industry firms, industry institutes and federations, further education establishments, key pan-industry bodies such as Rethinking Construction or Be, and those in key positions at the DTI, must openly acknowledge that the first task of any leader is to establish precisely where they are. Unless they can be certain about current levels of performance they cannot possibly set meaningful improvement targets.

This also requires leaders to agree which aspects of performance have the greatest potential to reduce the capital cost of construction. As labour and materials account for around 80–90% of the total initial capital cost of construction, and as the vast majority of this relates to work carried out by specialist suppliers (sub-contractors and manufacturers), it is obvious that the measurement of the effective utilisation of labour and materials by specialist suppliers should be given the highest priority.

Not only do construction contractors need to give top priority to measuring the effective utilisation of labour and materials by their specialist suppliers, they also need to give a high priority to the development of systems which ensure that the improvements in the utilisation of labour and materials are recorded and the savings so generated are captured and shared with the entire design and construction supply chain, including the end-user client, in the form of reduced tender prices for subsequent contracts and a reduction in the final account of the current contract.

If we go back to the four prime EFQM Excellence Model criteria listed at the start of this chapter, we can derive a series of top priority actions against each criterion:

❑ **Leadership.** The leaders cannot develop viable improvement goals unless they know precisely where they, and their suppliers, are at present in terms of every aspect of performance at project and strategic level. Unless they first measure the effective utilisation of labour and materials throughout the design and construction supply chain, they cannot know how much improvement is either possible or desirable and they cannot prioritise the key areas for improvement. In fact, without performance measurement at project and strategic level, leaders could well waste a great deal of money and resources trying to improve aspects of performance that are already effectively done.

❑ **Processes.** An organisation cannot improve the management and execution of the working practices of its own people, and those of the people within its suppliers, unless it first measures how effectively its existing working practices are performing. There is little point in assuming or guessing at current levels of effectiveness, especially where they relate to the performance of people in other firms, because the assumption or the guess is highly likely to be wildly inaccurate. As an example, a Chief Executive of an industry firm assessed the effective utilisation of labour within his firm to be around 85%. Unfortunately when the effectiveness was accurately and independently measured it came out at 20%. Improving the working practices within the design and construction supply chain first requires accurate performance measurement of all working practices at project and strategic level.

❑ **Customer results.** The firms in the design and construction supply chain should not merely assume that their constructed products are fully satisfying their end-user customers, especially in a situation where numerous end user-inspired reports are expressing considerable dissatisfaction. As in other sectors, customer satisfaction should be independently measured and the aspects that should be assessed with the greatest care are those that are of greatest importance to the competitive performance of the end user. The National Audit Office report *Modernising Construction* and the Confederation of Construction Clients *Charter Handbook* provide excellent pointers to what is really important to end users in the UK.

❑ **Key performance results.** An organisation can only know what performance it is really achieving if it measures *every* aspect of its performance. In the case of construction contractors, since 80–90% of the resources used on construction sites come from the specialist suppliers (sub-contractors and manufacturers), there is little point in only measuring the performance of the construction contractor's own people. The full picture is only possible if the construction contractor ensures the specialist suppliers are measuring their performance and are making the information available without manipulation to the construction contractor. Similarly, project managers, design consultants and quantity surveyors cannot give end-user clients any assurance about effective performance and the elimination of unnecessary costs at project level unless they have evidence from performance measurement from every firm in the design and construction supply chain.

As the EFQM Excellence Model has not been widely taken up within the construction industry, the Construction Group within the British Quality Foundation has been working with an organisation called BQC Performance Management Ltd and the Construction Best Practice Programme to adapt the EFQM criteria for easier usage within the construction industry. The resulting guidebook is entitled *The Construction Performance Driver – A health check for your business* (see Further reading) and provides an excellent tool for improvement. In its section on leadership and under the question 'Am I a good boss?' it states:

'You are the one who decides most things in your company. Not only what gets done, but how. Your employees see you

around all the time and your "fingers are in all pies". It is likely that if anyone is going to pick up habits, both good and bad, that it will be from you. Have you realised yet that the way you manage your people and projects has a significant impact on whether your business is a success or not?'

The leadership section then goes on to pose a series of questions that the Chief Executive (and senior managers) need to ask themselves, such as:

- ❑ *'1.0 Are you always looking for improved ways of doing things?*
- ❑ *1.1 Do your management systems and methods encourage continuous improvement?*
- ❑ *1.2 Do you have effective ways to keep your employees informed of your desires and intent for the business?*
- ❑ *1.5 Do you have a plan to ensure you stay close to, and understand, your key customers and partners?*
- ❑ *1.6 Do you cultivate partnership-style relationships with your customers and suppliers?*
- ❑ *1.7 To avoid assumptions and misunderstandings, is there some formality in customer and supplier relation-ships?*
- ❑ *1.14 Do you ensure that resources are made available to support business improvement priorities?'*

The Construction Performance Driver – A health check for your business is a must for any construction industry firm that is intent on improving its competitive edge and its profit margins. It is closely aligned with the radical reforms demanded by every report from Latham onwards, and people at the sharp end of the construction industry (namely those that form the BQF Construction Group) have specifically devised its language to ensure ease of usage. Its statements under each section heading are remarkably perceptive and apt, for instance under the question 'How do we do things?' it states:

'Ideally, you take your time, money, materials, ideas and expertise and put together the things your customers value. The way you do this is crucial. You may have developed your methods bit by bit over the years, never having the time to stand back and see it all in perspective. If you have a few big customers you no doubt spend a lot of energy just trying to keep them happy but perhaps not looking beyond the fulfilment of the next contract or project. There are many different ways of doing things but whatever your approach, it is continually working at improving the way you do things that could well hold the key to beating the competition or, in some cases, surviving at all.'

It follows from the above that effective leadership demands a deep and clear understanding of what needs to be done differently, why it needs to be done differently, what improvements in performance will result from doing things differently, and how both the current performance and the performance resulting from those improvements will be measured. The next two chapters endeavour to explain how measurement can radically improve performance at project and strategic level.

7 Performance Measurement at Project Level

Earlier chapters explained why construction industry firms need to measure the current performance of their supply-side design and construction project teams in order to know how well they are doing in the areas that represented the largest element of the total construction cost, namely the labour and materials element which constitutes 80–90% of the total construction cost.

To give an accurate picture, the first imperative is to measure performance at site level on a range of typical contracts in order to establish a baseline performance at a given point in time. Then it is essential to monitor the rate of improvement on all subsequent contracts in order to spot the sites where performance is particularly good (or particularly bad). In both cases the performance measurement system needs to discover the cause of the good (or the bad) performance so that the lessons learned can be exported to all contracts and the firm's rate of improvement reinforced.

The labour and materials element of the total construction cost is primarily provided through a series of specialist suppliers (sub-contractors, trades contractors or manufacturers), consequently it is imperative that these are persuaded of the benefit of measuring their performance and, most important, persuaded of the benefits of sharing the results with the other members of the design and construction supply-side team.

Experience has shown that because of the adversarial relationships that pervade the industry, the specialist suppliers are unlikely to be persuaded to share their results with other specialist suppliers or with the construction contractor, unless they have the security of long-term supply-side partnerships with those other firms, i.e. they are part of the 'Virtual Firm' described in an earlier chapter. This reality is not just found in the construction industry, it is common to other sectors such as manufacturing and retailing.

The persuasion is also likely to be more effective if the specialist supplier can see how measuring performance can directly improve and assure their profit margin and therefore their competitive ability. In Chapter 2, a chart (Fig. 2.1) was used to show how much unnecessary cost lies concealed within the total construction cost in the form of the ineffective utilisation of labour and materials. If specialist suppliers can be persuaded that any reduction in unnecessary cost will be shared fairly with them and that they will be allowed to use their share to improve their profit margin, they are more likely to be persuaded of the advantages of measuring their performance and of sharing the results of that measurement.

In a fragmented industry, where adversarial attitudes have grown over the years, there is a tendency for firms to be hypersensitive to any suggestion that their performance is less than ideal. As a consequence, performance measurement tends to be viewed as a hostile and intrusive audit and a covert attack on profit margins and on management skills. This attitude is probably the reason why the Building Research Establishment CALIBRE performance measurement system has not been actively taken up by supply-side firms. Where it has been used, it has generally been demanded and funded by the more astute demand-side clients who saw it as a very effective way of reducing the unnecessary cost caused by the ineffective utilisation of labour and materials and thus reducing the initial capital cost of their project. Those demand-side clients that formed the UK Construction Round Table were among those that saw the considerable benefit of utilising the CALIBRE performance measurement system as an effective improvement tool.

Whilst the Building Research Establishment CALIBRE performance measurement system has shown time and time again that it can be used as a very efficient means of measuring and driving up the effective utilisation of labour and materials, it is also true that the savings generated down at specialist supplier level (and the lower level of their sub-suppliers) often do not filter up to the main contractor or to the end-user client. This is generally because the contracts between the main contractor and the specialist suppliers (sub-contractors, trade contractors and manufacturers) do not require the use of performance measurement or the sharing of the savings generated by performance measure-ment. The following case history illustrates this reality very effectively.

Case history – Building Down Barriers pilot projects
The end-user client's agent, Defence Estates, insisted that the Building Research Establishment CALIBRE performance measurement system was to be used on one of the pilot projects.

As the main contractor was not persuaded that the cost of undertaking the CALIBRE performance measurement system (roughly £120 000) would be offset by the savings generated by the more effective utilisation of labour and materials, the end-user client (the army) agreed to meet the cost of CALIBRE because they were convinced that there would be far less disruption of site activities with CALIBRE in place, the required handover date would be met and the end users would not have the cost and inconvenience of the usual time over-runs that had beset other projects.

The main contractor's estimate of the effective utilisation of labour and materials was around the industry norm (between 30% and 40%) and the estimate of materials wastage was also around the industry norm (10–30%). As the end-user client was paying the full cost of the CALIBRE performance measurement system, it was used to its full effect with surveys of what every operative was doing on site every two hours. The survey data was immediately fed into the Building Research Establishment computer to be processed and the result was quickly sent back to site so that the site management team (including the specialist supplier's site management) could see what was not going well while it was still happening on site and they could still do something about it.

At the end of the contract, the effective utilisation of labour had risen from the main contractor's estimate of 30–40% up to an amazing 70%. Similarly, the materials wastage had reduced dramatically from the main contractor's estimate of 10–30% down to less than 2%.

Unfortunately, because of the way the specialist supplier sub-contracts (and their sub-contracts) had been set up, it was impossible for the very considerable savings to be captured and shared with those higher up the supply chain, especially with the main contractor and the end-user client (who paid for the CALIBRE performance measurement system with the hope that the client's share of the savings would cover the cost of CALIBRE).

Interestingly, I calculated that the savings that must have been generated from the use of CALIBRE would have exceeded the cost of the CALIBRE performance measurement system by a factor of at least 6 and possibly as high as 10.

The Building Down Barriers case history clearly demonstrates why any form of performance measurement must not be introduced on an individual project without the full, active and enthusiastic participation of every member of the design and construction supply chain, especially the specialist suppliers and the firms in their supply chains.

Everyone must fully understand the commercial benefits that will come to every firm in the supply chain (including the end-user client) from the use of performance measurement as the foundation of a continuous improvement system. All must have a common understanding of what is to be measured, how it is to be measured, who will have access to the results, how the performance measurement system is intended to drive up the effective utilisation of labour and materials, how the man-hours and materials savings will be captured and recorded, and how the savings will be shared with the total supply chain.

The Building Research Establishment CALIBRE productivity toolkit was developed by the Building Research Establishment Centre for Performance Improvement in Construction (CPIC) and provides a simple but effective tool that enables site processes to be clearly mapped and understood. CALIBRE provides a consistent and reliable way of identifying how much time is being spent on activities that directly add value to the construction and how much time is being wasted on non-added value activities. CALIBRE has been developed to enable firms to improve on-site performance on a day-by-day, project-by-project basis.

The approach utilised is for a CPIC trained observer (who could be a CPIC trained member of the project team) to tour the site at regular intervals with a hand-held computer and record the tasks being undertaken by each operative. The captured data is then quickly analysed to identify how much time is being spent on activities that directly add value to the construction and how much time is being wasted on non-added value work.

Performance measurement systems such as CALIBRE operate in real-time and enable the site team to quickly see the effects of any process changes that are introduced to improve the site productivity. Because such systems record data against standard codes, productivity and performance information can be compared on a day-by-day, site-by-site and supplier-by-supplier basis for any combination of work, area, task, activity or operative.

The use of performance measurement systems such as CALIBRE is not the only way of assessing how well the project supply chain firms are performing. The specialist suppliers' (sub-contractors, trades contractors and the firms in their own supply chains) operatives know only too well where things go wrong site after site after site. They are also the people most likely to come up with excellent ways of correcting the things that regularly go wrong and thus improving the performance of the entire supply-side design and construction team.

Case history – Building Down Barriers pilot projects

Although the Building Research Establishment CALIBRE productivity toolkit was used on the larger of the two pilot projects, the smaller project elected to improve productivity (the effective utilisation of labour and materials) by forging a high degree of openness, trust and collaboration between all the members of the design and construction supply-side team. The intention was to create an environment where individual operatives would open up and be honest about the magnitude, the nature and the causes of disruption and reworking, and the design consultants would be willing to listen to the operatives and use their experiences to improve the design.

This required the main contractor's project manager to have outstanding facilitation skills and also required the project manager and the design consultants to listen to the operatives and not to react negatively and aggressively when any cause of disruption and reworking was placed in their camp. Initially, both the architectural and the structural/civil engineering firms found this feedback from the operatives impossible to bear and refused to accept that any operative had sufficient knowledge or understanding of design to be able to criticise the work of a design professional, or to suggest ways of improving the design.

In fact, the situation proved impossible to resolve and eventually both firms withdrew from the project and were replaced by firms that agreed to actively participate in the two-way dialogue and to accept the operatives as equal team members. Interestingly, the selection system used by the supply-side to find replacements for the architectural and structural/civil engineering firms included the specialist suppliers because they insisted that only they could ensure that the replacement firms would treat the specialist suppliers as equal team members with a right to advise on design.

Once every team member had agreed to be open, honest and receptive about the extent and the causes of disruption and reworking, an enthusiastic collaborative learning culture quickly developed across the entire supply-side team. The operatives felt encouraged and able to open up about where they had traditionally been rendered less effective by actions beyond their control. They felt free to suggest the most likely causes of the disruption and reworking on past projects and they had sufficient confidence to warn where they thought aspects of design of the Building Down Barriers Pilot project might well cause disruption and reworking.

With the benefit of the project manager's outstanding facilitation skills, the specialist supplier's operatives and the design consultants were able to enter into a constructive dialogue to find the most effective way of improving the design to resolve the causes of reworking and disruption.

In the case of the steel fabrication specialist supplier, harnessing the skill and experience of the fabricators and the erectors to improve the design of the steel frame achieved major savings in the capital cost of the steel frame and a major improvement in their profit margin. In fact, they erected the steel frame 'right first time' with major improvements in the use of standard components.

In addition, they were convinced that the lessons they and the design consultants had learned on the pilot project meant they could take 15% off the capital cost of any subsequent steel frame **if** the design and construction team could stay together.

It is obvious from these two case histories that whilst performance measurement systems such as CALBRE are an excellent and standardised means of measuring and improving site productivity, the same improvement can be achieved through constructive dialogue between the design consultants and the specialist supplier's operatives.

But the constructive dialogue will only be possible if a culture of total openness, trust and collaboration can be established across the entire design and construction supply-side team.

However, the Building Down Barriers pilot project experience also suggests that the constructive dialogue will only work if the main contractor's project manager has excellent facilitation skills and is willing to put aside the usual 'stick and carrot' adversarial approach. In addition, both pilot projects also taught that performance measurement would only work if the site team had the full and overt support of their head office managers, who fully understood the purpose of the more collaborative way of working and were determined to use it to drive out unnecessary costs in the form of the ineffective utilisation of labour and materials.

The experience of the Building Down Barriers pilot projects suggests that not all firms (design consultants, main contractors and specialist suppliers) can live with the necessary degree of openness about what regularly goes wrong on site and what (or who) regularly causes the disruption and reworking. The Building Down Barriers pilot project experience also demonstrated very clearly that the responsibility for creating the open, trusting and collaborative culture rests firmly on the shoulders of the main contractor's project manager, but that the project manager must have the overt and active support of those in management positions directly above.

The project manager on site needed to have a very high degree of facilitation skills to calm the fears of anxious team members and to persuade them to be open and honest about what (from the specialist supplier's experience) aspects of the design might cause disruption and reworking, and then (from the design consultant's perspective) to have the humility to accept criticism and constructive advice from other team members.

The on-site project manager is key to the success of the whole improvement process. He or she must break down the traditionally fragmented and adversarial culture of the supply-side team and persuade them to see each other as equals in terms of knowledge, experience and skills. The professional members of the design team must recognise the value and the knowledge of the specialist supplier's operatives and must welcome their input into the design from the outset. In fact, the design professionals must ideally invite the specialist supplier's operatives into the value management workshops that tease out and clarify the full extent of the end user's functional requirements, so that those constructing the building intimately understand the detailed functional requirements that the building must deliver efficiently when complete.

It is imperative that the entire supply-side team recognise that most of the things that cause problems on site emanate from the design and are almost certainly locked into the design at a very early stage in the design process. As a consequence, the design professionals need to seek advice and input from the specialist supplier's operatives at the outset of design development.

It must be emphasised that to be of any real use the advice and input must come from those operatives who will be constructing the building in question and therefore have a vested interest in ensuring that they inject maximum build-ability into the design, so that they will be able to construct their particular construction activities 'right first time'. There is absolutely no point in the design professionals seeking advice and input from any old specialist supplier who has nothing to gain from the relevance or the quality of their advice and input.

Before any design development starts, an open, trusting and collaborative culture must be forged between all members of the design and construction supply chain so that the specialist supplier's operatives feel sufficiently confident of their status in the team to set out those things that cause disruption and reworking on sites with awful regularity. The culture must also persuade the design professionals to have sufficient regard for the knowledge of the specialist supplier's operatives that they truly listen to their advice and truly make use of their input into the design.

Time and time again, specialist suppliers' operatives have made clear to me that it is extremely rare for them to come across a cause of disruption or reworking that is unique to a specific site. The reality they find is that the same causes of disruption and reworking occur on site after site after site. Despite claims by design professionals that disruption and reworking would be avoided if they were given sufficient time to complete the working drawings, specialist suppliers' operatives insist that they have never come across a fully complete and perfect set of working drawings that do not conceal numerous causes of disruption and reworking. Even when they have been presented with a so-called complete and perfect set of working drawings, the reality that rapidly emerges on site is that the drawings are riddled with errors, omissions and co-ordination clashes that cause the usual level of disruption and reworking.

Creating a truly open, trusting, collaborative and mutually supportive culture where performance measurement in any form would be possible, will almost certainly require a radical change in the way that the supply-side firms contract with each other. Unless there are sufficient incentives built into the supply-side contractual relationships, it is unlikely that the design and construction team members will be fully and enthusiastically co-operative.

Why should the specialist supplier's operatives be open about the level of effectiveness of their on-site performance or their wastage of materials, unless they gain from their openness? Why should they hand over the benefit of their hard-won knowledge and experience unless they gain something in return? Why should they give their valuable time at the outset of design development unless they can be sure that they will be the ones that construct the design? Why should they go out of their way to help and support the design professionals during design development if they can only be certain of the current contract?

Before any of the specialist suppliers (sub-contractors, trades contractors or manufacturers) are willing to be open and honest about the causes of disruption and reworking, or to make their knowledge and experience available to the design professionals from the outset of design development, they will almost certainly demand the security of long-term, strategic supply-side partnerships from the construction contractor. Before any of the design professionals (architects, structural/civil engineers and mechanical/electrical engineers) are willing to accept criticism from the specialist suppliers and to welcome input from them into the design from the outset, they too will almost certainly demand the security of long-term, strategic supply-side partnerships from the construction contractor. Consequently, the construction contractor will need to urgently rethink the nature of the supplier relationships before any form of performance measurement can occur.

In Chapter 5 on the 'Virtual Firm' I explained in detail the nature of these long-term strategic supply-side partnerships and I said that partnering would deliver the greatest improvements in performance where it was the basis of the long-term strategic relationships between firms on the design and construction supply-side of the industry. I also said that this would necessitate a radical and profound change in the way the supply-side firms operate and this would in turn require their Chief Executives to fully understand the nature of the changes in working practices that must be put in place within their own firm and within the firms with which they do business.

The Chief Executives must measure their organisation's current performance (such as the effective utilisation of labour and materials) and that of their suppliers so that they have a firm basis from which to start the improvement process. They must then become fervent and committed champions for the changes in working practices, because only powerful and clear-sighted leadership from the Chief Executive can make those changes happen. This message applies particularly to the Chief Executives of construction contracting firms (main contractors) who, because of their traditionally key role in the supply-side team, are most likely to be the ones that need to make the first move with their relationships with their suppliers.

The incentives available to construction contractors that are most likely to be effective in persuading suppliers (design professionals, trades contractors, specialist contractors and manufacturers) to be open, honest, collaborative and enthusiastic contributors to the performance measurement and the improvement processes, are primarily non-financial in nature.

The first and most obvious incentive is for the construction contractor to agree and ring-fence suppliers' (design professionals, trades contractors, specialist contractors and manufacturers) profit margins at a reasonable level. This is perfectly possible if the goal of the performance measurement and improvement processes is to drive out unnecessary costs. The design and construction supply-side team (the 'Virtual Firm') ensure their competitive success by continuously reducing unnecessary costs, not by slashing profit margins to a level where firms become insolvent.

The second major incentive available to construction contractors to persuade suppliers to embrace performance measurement and continuous improvement as part of effective supply chain management, is a reasonably assured flow of work. Having to tender for every project costs a great deal of time and money; having to also base the tender on the lowest cost means that the supplier's business prospects are financially risky and uncertain.

If the construction contractor's relationship with its suppliers (design professionals, trades contractors, specialist contractors and manufacturers) was that of long-term strategic partnerships, and if those partnerships gave a reasonable assurance of a flow of work and ring-fenced profit margins, then the suppliers would have the confidence to be open about the degree to which on-site performance is affected by disruption and reworking, would be willing to share their knowledge and experience with their supply-side partners and would be willing to accept that all could play a vital part in the design development process.

Assuming that the supply-side design and construction firms have formed themselves into a 'Virtual Firm', the following action plan may help them to fit performance measurement into the design and construction process of individual projects and thus drive out the unnecessary costs caused by the ineffective utilisation of labour and materials. In situations where the architect and the design team are appointed well in advance of the construction team, or where the concept design is done within the client's organisation and imposed on the supply-side team, the action plan I have suggested will be impossible to apply without considerable adjustment.

Even if it proves possible to utilise a heavily adjusted form of the action plan, it is highly unlikely that the specialist suppliers will be available at an early enough stage in the design development to be able to influence the design sufficiently to take out much of the unnecessary costs.

SUPPLY-SIDE ACTION PLAN FOR THE INTRODUCTION OF PERFORMANCE MEASUREMENT AT PROJECT LEVEL

This plan suggests possible actions for the supply-side firms that are determined to use some form of performance measurement to eliminate the causes of disruption and reworking and thus eliminate unnecessary costs and convert them into a lower out-turn price to the end-user client and higher profit margins for the supply-side firms. The action plan is broken down into project stages since the necessary actions vary from design inception through to construction.

Although I have based the following action plan on the construction contractor or main contractor taking the lead in the formation of the long-term strategic supply-side partnerships, there is no reason why other members of the supply-side design and construction team should not take the lead.

I certainly know of at least one instance (from my involvement with prime contracting at Defence Estates) where a major quantity surveying consultancy realised that the primary requirement for taking the lead in the formation of the virtual supply chain that constituted a prime contracting team was project management skills. Since they had an abundance of such skills they decided they were as well equipped as any construction contractor to take the lead in the formation of the virtual supply chain for a major prime contract. Their assumption proved to be absolutely correct because they beat the construction contractor-led virtual supply chains in each stage of the assessment process and were awarded the prime contract.

Thus the following should be read bearing in mind that where I have said 'construction contractor' or 'main contractor' the action plans would work equally well if any member of the supply-side team elected to take the lead. In fact, in the world of supply chain management and supply-side integration, the terms 'construction contractor' and 'main contractor' may need to be replaced with the more logical term 'lead supplier'.

Project Level Action Plan Stage 1 – Strategic actions for the supply-side firms at pre-award stage:

As soon as a contract has been awarded by the end-user client to a design and construction supply-side team for the design and the construction of a new building, it is essential for the entire supply-side team to come together before design development is started to share their knowledge about the magnitude and nature of disruption and reworking on other recent and current sites.

The construction contractor and each of the specialist suppliers (sub-contractors, trades contractors and manufacturers) must be completely open about the degree to which their on-site performance is affected by disruption and reworking. This information must be based on measurement or on objectively assessed reality by the on-site operatives and must not be based on the optimistic assumptions of senior managers that have not been validated by the on-site operatives. Consequently it must be derived from a productivity measurement system such as CALIBRE or from facilitated workshops that encourage the specialist supplier's operatives to be totally open about the magnitude and the nature of disruption and reworking.

The design professionals in turn must listen carefully to the specialist suppliers and must be willing to acknowledge the validity of their concerns about deficiencies in the designs of recent or current projects and must be willing to accept that the specialist suppliers have the knowledge and the skills to advise on how designs could be improved to eliminate unnecessary costs.

The only way that this degree of co-operation and openness is likely to occur is if those firms that are likely to constitute the design and construction supply chain (especially the specialist suppliers) are already committed to working together within strategic supply-side partnerships.

This change in supplier relationships needs to be at a strategic level that is totally independent of individual projects and should be done in advance of any specific project. It is likely to be most effectively done if it is initiated by a construction contractor who is determined to deliver better value to end-user clients and to achieve much higher profit margins by eliminating unnecessary costs caused by the ineffective utilisation of labour and materials. Such a construction contractor would want to move away from giving the buying department of the firm, or the site agents, a free hand in the selection of suppliers (design professionals, sub-contractors, trades contractors and manufacturers) on the basis of the lowest price.

I frequently come across construction contractors that have a supplier (design professionals, sub-contractors, trades contractors and manufacturers) database in excess of 20 000 firms (the worst case was 90 000 firms). These databases have built up over the years because buying departments and site agents have been allowed almost total freedom to go to the market on every project in order to try and obtain the lowest price from each supplier.

In fact, I have frequently come across cases where the buying department have gone to the market twice on the same project in order to try and drive supplier prices down to the lowest possible level. They have first gone to the market when the tender is being put together, and they have then gone to the market again when the contract has been awarded to their firm in order to try and drive the suppliers' (design professionals, sub-contractors, trades contractors and manufacturers) prices even lower and thus boost the construction contractor's own profit margin at the expense of the suppliers' profit margins.

Unfortunately, this endless search for the lowest prices ignores the reality that the lowest price is rarely the best value. After all, in such a situation a wise specialist supplier is likely to hide a sizable contingency sum within the supposedly lowest price quoted to the main contractor in order to cover the disruption and reworking that will almost certainly be encountered on site in a situation where every supplier (design professionals, sub-contractors, trades contractors and manufacturers) has been driven down to the lowest possible price. The endless search for the lowest price also ignores what is really happening on site in terms of the effective utilisation of labour and materials, which frequently generates unnecessary costs that consume 30–40% of the total project cost.

Once a construction contractor decides to base its competitive success on the elimination of unnecessary costs, the way is open for a radically different approach to the selection of suppliers (design professionals, sub-contractors, trades contractors and manufacturers).

A continuous improvement process that is targeted at the elimination of unnecessary costs and the delivery to the end-user client of best whole-life value requires the construction contractor's relationship with its suppliers (design professionals, sub-contractors, trades contractors and manufacturers) to be based on long-term, strategic partnerships that embody the seven universal principles of supply chain management (see Chapters on the 'Virtual Firm'), especially the primary principle that advocates the construction contractor *'competes through superior underlying value'*.

This requires the construction contractor to examine its historic workload in terms of components, materials and trades to spot the patterns running through successive projects that suggest types of suppliers (design professionals, sub-contractors, trades contractors and manufacturers) and the number in each category, that would be necessary to sustain a similar workload trend in the future. Services are an easy area of commonality to spot since every building involves mechanical and electrical services in some form and there is very little variation from the norm since very few buildings have highly specialist services, such as air conditioning. Windows and steel frames are other obvious areas of commonality, since virtually every building has windows (only the material the windows are made from varies) and around 70% of buildings have steel frames.

Once patterns have been spotted it should be fairly simple for the construction contractor to work out how many suppliers (design professionals, sub-contractors, trades contractors and manufacturers) would be necessary to deal with the likely future workload (in terms of the anticipated turnover, the variety of buildings and facilities and workload peaks) in the geographical area covered by the construction contractor. The absolute minimum would probably be two in each category of supplier to cover the construction contractor's total geographical area. The two would be necessary to give a degree of flexibility and to allow benchmarking between suppliers in the same category. But the maximum might well be two or three suppliers in each category in each geographical region where a construction contractor has a wide geographical base and a large turnover.

The process by which the construction contractor selects the preferred specialist suppliers was covered in some depth in my book *Building Down Barriers – A guide to construction best practice* (see Further reading). Whilst it would be sensible to obtain a copy of the book to fully understand the context of the following extract, the key section of the book is reproduced below:

❑ '*Assess the current level of understanding of the three best practice standards and their two key differentiators and their six primary goals within the firms that form your current down-stream supply chain. Never make assumptions about what you think people know, always seek evidence to demonstrate what individuals really understand. Ensure the assessment is carried out consistently across your suppliers by the use of a simplified, easy to understand version of the three standards as the common evaluative criteria for all the assessments. Use an independent expert, who can demonstrate a comprehensive understanding of the three standards, to assist the assessment process. Ensure the assessment includes the attitudes of the Chief Executives and the senior managers, since the knowledgeable and enthusiastic support of the Chief Executive and the senior managers will be of paramount importance in any change process.*

❑ *Compare their current practices with the six primary goals, i.e. if they show no interest in forming long-term strategic supply chain relationships with you which embrace the* seven principles of supply chain management, if whole-life costing, value management and value engineering are not a well documented part of their normal practice, if they do not measure the effective utilisation of labour and materials, if they have no way of capturing and sharing efficiency gains with yourself or the rest of the supply chain, then their current practice is not the best practice of the three standards. They will therefore have difficulty in working with you to deliver the evidence required by those clients that have adopted the six primary goals and will lack the culture and practices necessary to form good strategic supply chain partnerships. Always ensure that this comparison is done objectively and never make assumptions or accept anyone's opinion as fact. It may be best to seek the assistance of an independent expert, who can demonstrate a comprehensive understanding of the three standards, to assist the comparative analysis.

❑ Consider what evidence is available to support the accuracy of the labour and materials elements of the prices tendered by each of your suppliers, i.e. is detailed process mapping used to work out the flow of work stream activities and the man-hours required for each activity? Is there a means of checking actual man-hours worked against those anticipated in the process mapping, and of apportioning cause for any deviation?

❑ Having assessed your current suppliers, decide on a short list of those that are most likely to support your drive for best practice. Choose from this short list those firms with which you wish to set up long-term strategic supply chain partnerships.

❑ If you have the assurance of your suppliers' commitment to you that can only come from within long-term strategic supply chain partnerships, consider how large your downstream supply chain needs to be, i.e. you will need to look back over your previous contracts and break down your typical workload into services, components and materials such that you can recognise patterns of commonality. Future workload trends are highly likely to follow historical patterns. As a consequence, you can use this commonality to set the down-stream supply chain partnerships that would give greatest value. You may only need two firms for a given component or service, unless there are particular geographical reasons for a greater number. The important consideration is that you should restrict the number of suppliers to those that your historical workload indicates you can provide with a reasonable flow of work. Without this down-stream workload assurance, it will not be worthwhile them making the investment necessary to embrace the best practice of the three standards. You also need to bear in mind the resource implications to yourself that will come from the need to work closely with each of your suppliers to develop and monitor mutually agreed improvement targets.

❑ You will also need to consider whether others, not normally in your supply chain, could add greater value by being included, i.e. you will need to look back and examine where errors in design have caused reworking and consider who needs to be included to eliminate such errors in the future.

❑ Discuss and agree improvement targets with each of your chosen supply chain partners that will enable them to deliver the six primary goals. Ensure everyone within their firm, and everyone within those firms with which they have an interface, fully understands what the target means and how their individual role will be affected.

❑ Discuss and agree Key Performance Indicators that they can use to accurately measure their firm's rate of improvement. These must dovetail with your Key Performance Indicators to ensure consistency in the overall improvement regime. For instance, set KPIs which require the increasing use of independent user surveys to measure end user satisfaction with functionality; which require the increasing production of, and increasing accuracy of, cost of ownership predictions by the design and construction teams for the end-user clients; which measure the rate of improvement in the on-site utilisation of labour and materials or which compare the construction team's predicted utilisation levels with the actual utilisation levels; which

> *measure when specialist suppliers actually become involved in design development, especially whether their skill, knowledge and experience is really being used to the greatest advantage by the designers; which measure the rate that your existing suppliers are brought into strategic supply chain partnerships; and which measure the speed at which single point procurement becomes the firm's preferred way of doing business.*
>
> ❑ *Set up arrangements to regularly monitor each supply chain partner's performance against their Key Performance Indicators. These should include an open-book approach to costing and a willingness to share innovations and improvements with other firms within your supply chain. They should also include sharing your own firm's performance with your suppliers and opening your own books to them in a totally trusting and open partnering relationship.*
>
> ❑ *Ensure all leaders, especially the head of each firm, become crusaders and champions for best practice. This necessitates their having a deep and consistent understanding of the six primary goals and a commitment to their delivery.'*

Once the construction contractor has decided the preferred list of specialist suppliers, the next step is to set up a workshop with each specialist supplier to decide and agree the mechanism by which the specialist supplier's performance in the elimination of unnecessary costs (the ineffective utilisation of labour and materials) will be measured.

The mechanism used for measuring the elimination of unnecessary costs ought to be consistent across all specialist suppliers or there will be no way of comparing (benchmarking) the results across all the preferred specialist suppliers. If a productivity measurement system such as CALIBRE is used for *every* specialist supplier, the standardised approach imposed by the measurement system itself will ensure that results can be easily compared and shared between all the preferred specialist suppliers.

If a self-assessment approach is adopted, such as was used in the smaller of the two Building Down Barriers pilot projects (see the case history earlier in this chapter), it would be advisable to adopt a standardised list of headings that covers the full range of avoidable delays and disruptions to on-site activities. BSRIA Technical Note TN 13/2002 *Site Productivity – 2002, A guide to the uptake of improvements* provides a very useful list of headings on its page 42, which could be adopted as a standardised approach. It would be sensible to obtain a copy of this Technical Note to understand the full context of the list and the reasons behind it, but the BSRIA TN 13/2002 headings themselves are reproduced here:

BSRIA TN 13/2002 list of avoidable delays and disruptions

- '(1) Off-site manufacturing error.
- (2) Permit or method statement issue.
- (3) Rework through design change.
- (4) Inclement weather.
- (5) Rework through installation error.
- (6) Waiting for instruction.
- (7) Obstructed work area.
- (8) Spatial clash/co-ordination problem.
- (9) Constraints from preceding work.
- (10) Drawing or specification issue.
- (11) Collecting and waiting for materials.
- (12) Collecting and waiting for plant/tools/equipment.
- (13) Late start/early finish/extended break.'

It should be borne in mind that whilst the preferred consultant design suppliers would not be directly involved in the measurement or the assessment of non-added value activities, they would need to have access to the results and would need to agree to work closely with the preferred specialist suppliers to eliminate those aspects that are caused by actions on the part of the consultant design suppliers, such as drawing, specification or co-ordination problems.

Once the measurement or assessment process has established the baseline of current performance for each preferred specialist supplier, an improvement regime that includes Key Performance Indicators will need to be agreed with each firm. Again, there needs to be a consistent approach across all preferred specialist suppliers so that all can see that the construction contractor is operating a level playing field.

The mechanisms by which the construction contractor can best partner with the preferred suppliers (design professionals, sub-contractors, trades contractors and manufacturers) were covered in depth in Chapters on the 'Virtual Firm' and will therefore not be repeated here. The key thing to remember is that supply-side strategic partnerships are likely to be most effective if they are based on the seven universal principles of supply chain management from the *Building Down Barriers Handbook of Supply Chain Management* published by CIRIA. These were explained in detail in Chapter 5 on the 'Virtual Firm' but very briefly are:

(1) Compete through superior underlying value.
(2) Define client values.
(3) Establish supplier relationships.
(4) Integrate project activities.
(5) Manage costs collaboratively.
(6) Develop continuous improvement.
(7) Mobilise and develop people.

The construction contractor's selection of the suppliers providing the design professionals is very similar to that suggested above for specialist suppliers. The main exception is that there would only be a limited requirement to measure the effective utilisation of labour by the design firms. Again, the end result of the selection process should be a series of supply-side partnerships with firms of architects, civil/structural engineers, mechanical/electrical engineers and quantity surveyors that are sufficient to cover the anticipated future workload.

Needless to say, it is imperative to ensure that every partnering firm of architects, civil/structural engineers, mechanical/electrical engineers and quantity surveyors gives an undertaking that they will treat the construction contractor's specialist supplier partners as equal team members, will always listen to their concerns about the buildability of designs and will always listen and utilise their advice on design improvements.

Project Level Action Plan Stage 2 – Actions for the supply-side firms at the project design stage:

As early as possible in the design development stage the full design and construction supply chain team need to come together to assess if there are any aspects of the concept design that might cause delay or disruption on site.

Such an assessment would draw heavily on the knowledge and experience of the specialist suppliers (sub-contractors, trades contractors and manufacturers) from previous similar projects and in doing so would obviously use the evidence of disruption and delays from the 'Strategic actions for the supply-side firms at pre-award stage' in Action Plan Stage 1.

Accordingly it would be sensible to base the assessment on the BSRIA TN 13/2002 *Site Productivity – 2002, A guide to the uptake of improvements* (see Further reading) list of headings:

- ❑ Degree of standardisation of components and materials.
- ❑ Congested work areas.
- ❑ Spatial clashes/co-ordination problems.
- ❑ Specification problems.
- ❑ Delivery of materials.
- ❑ Availability of plant/tools/equipment.
- ❑ Potential for off-site manufacture.
- ❑ Sequencing of construction activities.

The above list is merely indicative and ought to be refined in consultation with the specialist suppliers in the project supply chain team.

The purpose of the assessment exercise at this stage is to assist the design team to ensure the developing design takes full account of the knowledge and experience of the specialist suppliers and thus lends itself to the achievement of a 'right first time' culture on site. The output from the assessment ought to be a confirmation from each specialist supplier that every element of the design can be constructed 'right first time' and will cause far less disruption and reworking than on previous projects. It might prove possible for the design and construction supply-side team to agree a target for the unnecessary costs generated by the ineffective utilisation of labour and materials.

The final actions at the project design stage are to agree the following:

(1) The system that will be used on site to measure the actual performance in the effective utilisation of labour and materials.

(2) The individual responsibilities for every aspect of design and construction, especially the interfaces, since most problems on site stem from a lack of clarity over who is designing what, or who is constructing what. These responsibilities need to be recorded in a comprehensive interface register that will need to be refined and updated as the project proceeds.

(3) The method by which savings in unnecessary costs will be captured and recorded by each specialist supplier and by the construction contractor. This must also include agreement on the mechanism by which the information will be promulgated to, and shared with, every firm that makes up the entire design and construction supply chain team.

(4) The method by which savings in unnecessary costs will be shared across the entire design and construction

supply chain team, including the method by which they will be shared with the end-user client.

(5) The timescale and the process by which any additional member of the design and construction supply chain team will be appointed. This ought ideally to be restricted to highly specialist firms that are required to deal with any unique aspects of the project.

Project Level Action Plan Stage 3 – Actions for the supply-side firms at the project construction stage:

Immediately before any construction work starts on site it is imperative that the full supply-side design and construction team re-confirm the process by which each specialist supplier (sub-contractors, trades contractors and manufacturers) will measure their effective utilisation of labour and materials.

This ought to include the method of induction that will be used to ensure that all the site operatives of every firm involved in the construction process have a common understanding of what is to be measured, why the measurement is necessary and how the savings are to be shared. The induction sessions ought to be site based, ought to include a mixture of operatives from the various specialist suppliers and ought to be done on a regular basis that reflects the arrival of operatives on site who are unfamiliar with the performance measurement process (say every Monday).

Before any construction activities start it is also important to ensure that the mechanism by which savings in the unnecessary costs will be captured, recorded, communicated and shared across the supply chain is fully understood by the specialist suppliers and their own supply chains.

It is also important for the specialist suppliers to re-examine the drawings and specification to assess if there remain any aspects of the design or specification that might cause delay or disruption on site. The output from this exercise ought ideally to be confirmation from each specialist supplier that nothing exists on the drawings or in the specification that would inhibit their ability to deal with their aspect of the construction activities 'right first time' with the maximum effective utilisation of labour and materials.

It would be sensible for the construction contractor's project manager to block any start being made to construction works on site, or to off-site prefabrication, until every specialist supplier (sub-contractors, trades contractors and manufacturers) had confirmed their satisfaction with the drawings and specification. This ought to include confirmation from each of the design consultants (architects, civil/structural engineers and mechanical/electrical engineers) that they had incorporated all the buildability improvements to the drawings and specification that had been agreed with the specialist suppliers.

Finally, the firms making up the design and construction supply-side team ought to produce an interface register in which the responsibility for every physical junction or interface is recorded, so that everyone knows with absolute certainty who is dealing with every critical aspect of the construction. The importance of the interface register cannot be over-emphasised because problems, in both the initial construction and in the long-term operation of the building or facility, always occur where two components or two materials come together, i.e. windows never leak through the glass, they always leak where the glass joins the frame or where the frame joins the wall.

When the design and construction supply-side team are assessing the drawings and specification it would be sensible to use the same list that was suggested above in the section dealing with the project design stage.

The final action prior to the start of site activities is to reconfirm the following:

(1) The system that will be used on site to measure the actual performance in the effective utilisation of labour and materials.

(2) The individual responsibilities for every aspect of design and construction, especially the interfaces, since most problems on site stem from a lack of clarity over who is designing what, or who is constructing what. These responsibilities need to be recorded in a comprehensive interface register that will need to be refined and updated as the project proceeds.

(3) The method by which savings in unnecessary costs will be captured and recorded by each specialist supplier and by the construction contractor. This must also include agreement on the mechanism by which the information will be promulgated to, and shared with, every firm that makes up the entire design and construction supply chain team.

(4) The method by which savings in unnecessary costs will be shared across the entire design and construction supply chain team, including the method by which they will be shared with the end-user client.

(5) The timescale and the process by which any additional member of the design and construction supply chain team will be appointed. This ought ideally to be restricted to highly specialist firms that are required to deal with any unique aspects of the project.

Project Level Action Plan Stage 4 – Actions immediately subsequent to handover of the completed building:

Performance measurement at project level does not end at the completion of the construction stage.

The Institute of Civil Engineers' excellent guide to value management emphasises the need to 'Plan, Do and Review' if you wish to improve your performance. Unless you measure or assess the effectiveness by which your current performance delivers a product or a service that matches the expectations of the end-user client, you cannot improve.

You first need to carefully plan what you intend to do (especially if the actions are as complex and inter-related as they are in the design and construction process), you then have to carry out the pre-planned actions and record what happened so that it will be possible to compare the planned actions with the actual actions during the review stage. Finally, in order to learn from the experience and improve your performance it is essential to review what actually happened, i.e. what went better than you planned, what went to plan and what went worse than you planned and required crisis management.

If the supply-side firms at project level are serious about continuously improving their performance, they must review their performance immediately the project is complete and handed over to the end-user client. As the saying goes 'If you don't learn from your mistakes, you will be doomed to repeat them'.

The post-completion review needs to be a collaborative exercise involving all the supply-side firms that formed the total design and construction supply chain, since any deviations from the planned performance of an individual firm would have had ramifications for other supply-side firms, or may have been caused by deviations from planned performance by another supply-side firm. In a supply chain as complex as that in the construction industry, working out how to avoid the deviations from planned performance on the next project is best done by the whole supply chain, so that all can better understand what went adrift in the complex inter-dependencies that exist in the supply-side design and construction team.

To provide effective feedback to the continuous improvement process the post-completion review must compare planned performance with actual performance and must list all the causes where the actual performance deviates from the planned performance. The result of the review must be communicated to every member of the supply-side design and construction team, so it ought to be recorded in a concise report and issued to every supply-side firm (including those in the specialist supplier's own supply chains).

The post-completion review report can also be used as a very effective marketing tool where end-user clients are increasingly demanding evidence of improved performance and are no longer prepared to accept unsupported assurances from potential suppliers of design and construction services.

The post-completion review report proves that the supply-side firms involved are learning organisations that are intent on continuously improving their performance. It also proves they are working collaboratively as an integrated supply-side design and construction team and are measuring their performance in order to learn from their mistakes and thus drive out unnecessary costs so that they can offer their end-user clients the best whole-life quality for the lowest optimum whole-life cost.

In Chapter 2, 'The Unchanged Customer Demand', I reminded readers that the adoption of the principles in the Confederation of Construction Clients *Charter Handbook* had been set as an industry target by the Strategic Forum for Construction report *Accelerating Change*, namely:

'By the end of 2004 20% of construction projects by value should be undertaken by integrated teams and supply chains; and, 20% of client activity by value should embrace the principles of the Clients' Charter. By the end of 2007 both these figures should rise to 50%.'

In view of this target, the post-completion review should most sensibly assess the out-turn performance of the supply-side design and construction team under headings that reflect the key improvements demanded by the *Charter Handbook*:

- ❑ major reductions in whole-life costs
- ❑ substantial improvements in functional efficiency
- ❑ a quality environment for end users
- ❑ reduced construction time
- ❑ improved predictability on budget and time
- ❑ reduced defects on handover and during use
- ❑ elimination of inefficiency and waste in the design and construction process

It is fairly obvious that several of the above improvements cannot be properly assessed immediately after handover and will need to be revisited when the end users have been in occupation, and the building has been in use, for at least twelve months (i.e. it needs to have passed through at least one winter's heating season and one summer's cooling needs). It will therefore be necessary to carry out a final post-completion review after about 12–18 months of the building's usage by the end user.

Project Level Action Plan Stage 5 – Actions 12–18 months after handover of the completed building:

Several of the key end-user concerns about improved performance from the construction industry relate to the performance of the building or facility when it is being used by the intended end users. The *Charter Handbook* has listed these long-term operating improvements – see first four items above.

The USA report that set their National Construction Goals in 1996 made the very telling point that if the salaries of the occupants are taken into account, the running costs for a typical office block over a single year equal the capital cost of construction. As a consequence (the report goes on to strongly emphasise), the running costs over the design life of the building are of far greater importance than the capital costs of construction. In fact, the USA report stated:

'The primary value of building comes from the productivity of the occupants, which depends on the capability of the building to meet user needs throughout its useful life'.

Because of these commercially sound end-user concerns for radical improvements in the operational performance of the building or facility, there is a need for the supply-side design and construction team to follow up the performance of its completed buildings during the second year of operation. Undertaking a post-completion review during the second year of operation also has the advantage of giving the building and the end-users a chance to bed down and overcome any initial teething problems that may have come from unfamiliarity with a new building and its dynamic systems, such as the heating system, the mechanical ventilation system and the electrical system.

The follow-up post-completion review needs to separate subjective opinion from fact, both on the part of the end-users and the supply-side team that designed and constructed the building or the facility. As a consequence, it would be best done through a carefully structured and standardised questionnaire that ensures all aspects are covered and allows the results to be compared for every building that the supply-side design and construction team has delivered. This comparison is essential if the rate of improvement is to be measured building-by-building and year-by-year.

Suggested topics for the carefully structured questionnaire are:

❏ **Whole-life costs.** Whilst the second year's figures are very early in the life of the building, it should be possible to compare actual maintenance and running costs with the predicted costs (i.e. How do the actual fuel bills compare with the predicted fuel bills over the first winter? How do the actual hours worked on maintaining the building compare with the predicted hours?). If the whole-life cost prediction has been done properly in accordance with the confederation of Construction Clients' guide *Whole Life Costing – A Client's Guide* (see Further reading) or with the benefit of advice from industry experts such as the Building Performance Group (who are linked to the Housing Association Property Mutual insurance company), it should be possible to compare actual costs with predicted costs for the first 12–18 months operation of the building.

❏ **Functional efficiency.** With advice from the key end users it should be possible to put together a detailed questionnaire that draws out the truth about the real functional performance of the building (i.e. Are the end users able to do every functional activity captured by the value planning/value management process at the start of design development and recorded in the detailed project brief? Are all the end users' functional activities achieved with maximum efficiency? Does the design of the building assist the end users' to achieve maximum functional efficiency?).

❏ **Quality environment for end-users.** The above questionnaire about functional efficiency devised with advice from key end users should also be able to examine the level of delight the building has engendered in those that use it, including any visitors to the building (i.e. In the case of a hospital, has the quality of the building's internal and external environment improved the morale of the doctors and nurses? Is the well-being of the patients improved by the internal environment of the building and has this led to a reduction in the time the patients spend in hospital?).

❏ **Reduced defects.** Those responsible for the maintenance of the building should be able to advise if any defects have arisen since the building was completed. The end users also need to be asked about defects in the functional performance of the building since the UK *Charter Handbook* and the USA National Construction Goals make clear that the efficient functional performance of the building over its design life is a key commercial concern of end users. Obviously, any post-completion major defects will have been brought to the attention of the supply-side team, but the post-completion review ought to be looking much deeper and asking questions about minor defects in the performance of the building that are likely to have caused irritation to the end users (and thus damaged their morale and their delight in the building's performance).

Finally, the above action plan reviews ought to bear in mind that the EFQM Excellence Model strongly advises that a carefully structured and systematic bottom-up approach to self-assessment is most likely to reveal the truth about what is really happening at the sharp end of any business. Any other approach could well be misleading and superficial (particularly if it is through the rose-tinted spectacles of senior managers) and the results would therefore be of dubious benefit if used as the basis of a continuous improvement programme.

At all times during the measurement of performance at project level the supply-side design and construction team should remember the UK Department of Trade and Industry adage:

'If you don't know how well you are doing, how do you know you are doing well?'

The supply-side design and construction team should also remember the statement in the UK British Quality Foundation sponsored guide *The Construction Performance Driver – A health check for your business* that was produced for the Construction Best Practice Programme. In section 6 on customer results it poses the question:

'Are your customers getting what they want and need from you?'

It then goes on to make the following statement:

'The only way to be sure that your customers keep coming back for more, and that they tell others about how good you are, is to provide the products and the services they really want at the price they want to pay and at the time they want. Finding out what they really think about you could well be the most important thing you ever do. Perhaps the next most important thing is to find out what your customers think of your competitors. And, by the way, it helps to look at those companies who do it really well – you might get some good ideas.'

8 Performance Measurement at Strategic Level

At the end of the last chapter I said that industry firms ought to bear in mind that the EFQM Excellence Model strongly advises that a carefully structured and systematic bottom-up approach to self-assessment is most likely to reveal the truth about what is really happening at the sharp end of any business. Any other approach could well be misleading and superficial (particularly if it is through the rose-tinted spectacles of senior managers) and the results would therefore be of dubious benefit if used as the basis of a continuous improvement programme. At all times during the measurement of performance at project level the supply-side design and construction team should remember the UK Department of Trade and Industry adage–see page 128.

In this chapter I have moved up from project level to the strategic level within the supply-side firms that over-arch the individual projects, and I have endeavoured to explain why and how performance measurement should become an intrinsic part of the way every supply-side firm (design consultants, construction contractors, specialist/trades contractors and manufacturers) runs its business at the strategic level that over-arches and supports the individual projects. After all, there is very little point in a supply-side firm demanding that its operatives use some form of performance measurement to prove they are performing at maximum efficiency at project or site level, if the firm is not prepared to demand the same evidence of performance from all other employees, including senior management. As we all know, the 'Do as I do' argument is always far more powerful than the 'Do as I say' argument.

On the demand side of the industry, it is particularly important that the end-user clients appreciate the added impetus they could give to the reforms they are demanding of supply-side firms if they took the lead by assessing their own performance. As I said above, there is no more powerful argument than being able to say *'Don't just do what I say, do what I do!'*.

The power of being able to persuade others to do what you want by first doing it yourself was very well illustrated by the following salutary lesson I learned on Building Down Barriers.

> **Case history – Building Down Barriers and Defence Estates**
>
> I well remember the heated arguments in the early days of the Building Down Barriers project, when I was endeavouring to persuade the two pilot project teams to accept that their performance could be radically improved and that embracing the EFQM Excellence Model within their firms might be a good way of driving forward radical improvement.
>
> Their response to me was that if I, as a member of the senior management team at Defence Estates, was insisting that the supply-side firms were performing inefficiently and ought to embrace the EFQM Excellence Model as a means of assessing current performance, why was I not prepared to listen to their concerns about inefficiencies in the Defence Estates' procurement process and why was Defence Estates not prepared to embrace the EFQM Excellence Model? They said if I believed the EFQM Excellence Model was good enough for them to use it ought to be good enough for Defence Estates to use! They said that until Defence Estates

was prepared to validate the value of the Excellence Model by using it itself, they would doubt the real value of the Excellence Model.

Very clearly, the power of my arguments was severely weakened by my organisation's apparent refusal to do the things we were advocating others to do.

Interestingly, when the new Chief Executive arrived at Defence Estates a little later he immediately insisted that the organisation embrace the EFQM Excellence Model because he believed it was the only way that we could objectively measure our true effectiveness and therefore be forced to face up to reality instead of arrogantly deceiving ourselves with bloated assumptions about our effectiveness.

He firmly believed that any form of improvement process had to start by knowing how well you were really doing through objective and structured performance measurement. Needless to say, my standing with the Building Down Barriers pilot project teams rose considerably once they learned Defence Estates was finally doing what it advocated others to do.

If a UK construction industry firm intends to radically improve its performance, as the many reports over the last 70 years have demanded, it needs to start by being totally honest with itself about how well it, and its supply chain, is really performing in the utilisation of labour and materials and in the in-use functional and operational effectiveness of its buildings and facilities. It can only do this if it measures its own performance, and that of its supply chain, at all levels and then uses the evidence to put in place an improvement process that enables lessons learned on one project or site to be quickly transferred to all other projects or sites, or enables the lessons learned in each of the various off-site support sections to be quickly transferred to any other similar sections within the firm and thus improve the overall performance of the firm.

Unless supply-side firms use performance measurement to reveal the truth about their real performance they risk falling into the trap of self-delusion described in the following case history that I mentioned earlier in the book.

> ### Case history – Major specialist supplier
> The Chief Executive of a major specialist supplier was arguing with me about the effective utilisation of labour within the construction industry and was voicing a strongly held belief that the industry generally performed very efficiently. When asked how well his own firm performed he insisted that it was beyond reproach and achieved very high levels of effectiveness in its utilisation of labour.
>
> It then transpired that he based this assumption solely on the fact that there was never any labour in the yard. He had never measured how effectively the labour was utilised on site, he had no idea how much time was consumed by reworking due to errors on the drawings or defective workmanship, late deliveries of materials, congested or cluttered site conditions, poor programming of site activities, etc.
>
> Worst of all, he could see little to be gained from measuring the effective utilisation of labour on site because he could not see how he could gain commercial benefit from the expense of measuring what was actually happening on site or knowing precisely and accurately how much unnecessary cost was buried within his underlying labour and material costs.

Measuring and improving performance would obviously be best started at project or site level, because this can most quickly deliver hard evidence of early gains. These can then be used to persuade those involved with other projects or sites to embrace the culture and the processes that delivered those improvements. However, this sharing of lessons learned between project or site teams is not easy to achieve in practice where both the personalities and the firms making up the different project or site teams can (and generally do) vary widely because of the way they are selected for each project by the end-user client and the construction contractor.

If a supply-side firm intends to develop a continuous improvement process whereby the lessons learned on one project or site are to be exported to other projects or sites, it will obviously require a high degree of commonality between the various supply-side project or site teams. The continuous improvement process will obviously be most effective where the same supply-side firms are involved in the other projects or sites, and it will be even more effective where the same personalities are involved in successive projects or sites and can therefore continue to work together to apply the lessons they learned together in order to build on the earlier improvements.

The need to keep design and construction teams (which must include the specialist suppliers) together over a series of projects was emphasised in the *Accelerating Change* report, which said:

'Clients need a construction industry that is efficient. An industry that works in a "joined up" manner, where integrated teams move from project to project, learning as they go, driving out waste and embracing a culture of continuous improvement.'

As I explained in Chapter 5 dealing with the 'Virtual Firm', this need to keep supply-side teams together in order to continuously improve performance requires everyone in every supply-side firm to view buildings and facilities in a totally different light and to recognise the commonalities in components, materials and processes that pervade virtually all buildings and facilities. Once this commonality is recognised, the way is open for the same supply-side firms to work together in long-term strategic supply-side partnerships that will facilitate the continuous improvement of the design and construction process from project to project, regardless of which end-user client the supply-side team of firms are working with.

Recognising the need for performance measurement is not easy (as I showed with the Defence Estates example above) and will require senior managers, including Chief Executives, to have the courage (and the humility) to accept that their firm may be less effective than they have always maintained. As I showed above, it is very easy to be lulled into believing that a busy site or an empty yard means that everyone is occupied effectively and productively all the time.

Those at senior management level within supply-side firms frequently deceive themselves about performance at site level because they overlook the reality that for much of the time operatives are busy at reworking activities caused by such things as errors or omissions in the drawings, or by poor pre-planning, or by defective components or materials. They also overlook the reality that operatives are regularly disrupted by having to switch from activity to activity around the site (or between sites) because they are unable to complete any one activity because of poor pre-planning, cluttered sites and delays in the delivery of components or materials.

The reality of the chaos that exists on virtually every site was brought home to me very clearly when I was shown the results of various CALIBRE productivity measurement exercises. Time and time again, those construction activities that had been programmed to be completed in two or three visits to site had taken many more visits to site before they were finally complete. These numerous, unplanned returns to site had delayed the completion date, inflated the costs and wiped out profit margins on almost every project.

Performance measurement at site level is a useful tool for cutting through all the misconceptions about performance and is an excellent tool for accurately establishing the true level of effectiveness in the utilisation of labour and materials. Most importantly, it also provides the operatives on site with a very effective means of feeding back to senior managers the reality they have to deal with on site after site after site, especially the high level of commonality in the problems that arise on site after site with awful regularity.

In fact, in the case of Defence Estates, once the first EFQM self-assessment had been completed by those at the sharp end of the business and the report for the Executive Board on the outcome of the exercise had been made available to everyone within the organisation, those at the sharp end said that the best thing about the exercise was that the Executive Board could no longer ignore what those at the sharp end had been saying for years. Those at the sharp end said that the EFQM self-assessment finally stopped senior managers deluding themselves about what was really happening down at the sharp end and forced them to face up to reality and deal with the real operational problems. The EFQM Excellence Model had given those at the sharp end of Defence Estates a powerful voice that had to be listened to by the Executive Board, consequently those at the sharp end felt empowered by the Model.

A reality of the continuous improvement process that should not be ignored is that unless the leading firm (the main contractor) has a close, long-term relationship with the firms in the supply chain, it is almost impossible to set targets that relate to the elimination of unnecessary costs, or to major improvements in whole-life cost and quality. This is because 80% or more of those costs and virtually all aspects of quality are the responsibility of firms in the supply chain that are commercially independent of the leading firm or main contractor and over which the leading firm or main contractor has very little real direct control. This situation is made worse because it is impossible to totally eliminate all unnecessary costs or to radically improve whole-life cost and quality on a single project.

Radical change demands a long-term improvement process where small, incremental improvements are made on an individual project and then the lessons learned by the design and construction team (which must include the specialist suppliers) are applied and further refined and improved on subsequent projects. This continuous improvement process would make very little headway if the firms that formed the design and construction supply chain team on the first project were split up immediately the project was completed, and a new design and construction supply chain team were put together for the next project and for every subsequent project.

It therefore follows that for the continuous improvement process to deliver the greatest improvements, it must start with the formation of a 'Virtual Firm', i.e. a long-term supply-side partnership that includes the full range of design and construction firms necessary to cover all the skills required by the low-rise, low-tech buildings that form the bulk of construction activities. This enables the supply-side design and construction team to stay together over a series of projects (either for the same end-user client, or for different end-user clients) and thus further improve and refine the lessons they learn from project to project.

The security engendered by these long-term strategic supply-side partnerships also creates a mutually supportive and collaborative culture that enables the supply-side design and construction firms to act as a true team. They have the confidence to be totally open with each other about the lessons learned on each successive project (especially what went wrong) and can agree how each supply-side firm could best improve its particular aspect of the design and construction process so that the supply-side design and construction team can further improve their performance on the next and every subsequent project.

This supply-side partnering approach to continuous improvement also enables individual supply-side firms to become more competitive when approached by customers (demand-side customers as well as supply-side customers) for their services independent of the 'Virtual Firm'. This is because they can make use of the improvements to their performance that they have developed through their collaboration with their supply-side partners in the 'Virtual Firm'.

Thus every supply-side firm in the 'Virtual Firm' will inevitably become more competitive, whether operating within the 'Virtual Firm' or operating independent of the 'Virtual Firm'.

The following case history from the retail sector illustrates the above point extremely well.

Case history – Supermarket supplier

I was discussing the concept of continuous performance improvement within effective supply chain management with the Managing Director of a supplier to a major supermarket chain. He said that his firm was one of two preferred suppliers of a particular product for a major supermarket chain.

The terms of his contract with the supermarket required both suppliers to meet together with a representative of the supermarket every six months. Each supplier then had to be entirely open about any improvements they had made to both process (utilisation of labour) and to product (technical innovations) over the preceding six months. They also had to be entirely open about any problems they had encountered with their performance over the preceding six months.

The purpose of these regular six-monthly sessions was to ensure that each supplier could learn from the experiences of the other supplier. Each was expected to import process or product improvements from the other and each was expected (with the active support and help of the supermarket representative) to help the other solve any outstanding process or product problems.

When I voiced concerns that giving away process or product improvements in this way could be commercially damaging to each supplier, the Managing Director said that the opposite was the case.

He pointed out that in accordance with good business practice his firm had a wide customer base and only supplied some of its products to the major supermarket chain and the remainder went to other customers. In the wider market, his firm had many competitors other than the second preferred supplier to the major supermarket chain and his firm was rarely in competition with the second preferred supplier.

But because the two preferred suppliers were required to share their process and product improvements, each was more competitive in the wider market and each was able to improve their share of the wider market and their profitability when selling to that wider market.

Consequently, he was an enthusiastic supporter of the supermarket's requirement that its preferred suppliers must accept a totally open-book approach to process and product improvements. As far as he could see, the open-book approach could only become commercially risky to his firm if the two preferred suppliers dominated the entire market and were always in direct competition with each other for all their customers. Since he could see no possibility of such a duopoly situation ever arising, the supermarket's requirement that preferred suppliers share their process and product improvements was always going to be commercially beneficial to the individual preferred suppliers.

Interestingly, he said that his preferred supplier contract did have a stick as well as a carrot. If either of the two preferred suppliers refused to be open about their process or product improvements, or refused to import improvements from the other preferred supplier, their contract could be terminated. The reason for this was that the supermarket believed that continuous improvement was an essential part of their competitive success and was the main reason why they kept their position as a market leader. It was therefore essential that each preferred supplier was also an enthusiastic and committed supporter of the total continuous improvement process.

The above case history applies equally well to the construction industry, since the supply-side firms in the 'Virtual Firm' equate to the supermarket's preferred suppliers, and the total construction market is of a size to ensure that those supply-side firms will always have customers beyond those they are dealing with through the 'Virtual Firm'. As a consequence, sharing process or product improvements within the 'Virtual Firm' will always mean that the supply-side firms within the 'Virtual Firm' are more competitive and more profitable whether they are operating inside or outside the 'Virtual Firm'.

To illustrate this point, if a construction contractor develops a 'Virtual Firm' of preferred suppliers (design consultants, specialist/trade suppliers and manufacturers), that 'Virtual Firm' will contain at least two steel fabrication firms. Those steel fabrication firms will still supply steel components to other construction contractors and in that wider market they will always be in competition with many other steel fabrication firms. It therefore follows that the sharing of process and product improvements between the two or three preferred steel fabrication suppliers within the 'Virtual Firm' will make each preferred steel fabrication supplier more competitive in the wider market.

Once the 'Virtual Firm' is in being, those supply-side firms involved can start to work together at improving their performance at several levels. At project level, the individuals from the supply-side firms work collaboratively to drive out unnecessary costs, drive up whole-life quality and performance and drive down whole-life costs. The individuals then continuously feedback to their parent firms the process and product improvements they have made at project level so that the firms can import the improvements and ensure that these become the way they do their business for all their customers on all their projects.

In addition, the project teams will inevitably throw up process or product problems that cannot be solved at project level and require the supply-side firms in the 'Virtual Firm' to collaborate at strategic level to solve the problem. A possible example of this might be a problem caused by conflict between the steel frame and ductwork that only came to light on site and which was believed to stem from the timing of the steel fabrication firm's involvement in design development. To avoid the problem repeating on other projects, the architectural consultancy, the structural engineering consultancy, the main construction contractor, the mechanical engineering consultancy, the ductwork firm and the steel fabrication firm need to collaborate in order to agree the ideal time at which the steel fabrication firm ought to be involved in design development on subsequent projects.

Even before any work on forming the 'Virtual Firm' is started, it is imperative to ensure that the performance being measured and improved is fully aligned with those aspects of performance that are of concern to the eventual customers of the constructed products of construction industry firms, namely the demand-side end users who will occupy and use the buildings or facilities.

Throughout this book I have continually emphasised the importance of supply-side firms understanding precisely what aspects of their performance are of concern to end-user clients. *The Construction Performance Driver – A health check for your business* also picks up this need for construction industry supply-side firms to be certain about what aspects of their performance are important to the end-user customers.

There is little point in assuming that lowest capital cost is all important to the end-user customers if the reality is that they see lowest whole-life cost as far more important than lowest capital cost and are frustrated because they are not provided with an accurate whole-life prediction to insert into their long-term business plan. This is particularly the case in buildings, where all too often the consequence of driving down the lowest capital cost is that the whole-life cost increases because cheap, poor quality, low durability materials and components have been used or insufficient time has been allowed to develop and co-ordinate the design.

One of the priorities in any drive for improvement is to be certain that everyone is using the same language and has the same understanding of the various terms that are being used. This need for clarity in terminology and a very definite link between the supply-side's terminology and the terminology used by the demand-side end users is essential if both sides are to head in the same direction. The language used by the supply-side firms in their drive for improvement must be validated against key end-user publications (such as *Modernising Construction* or the *Charter Handbook*) if they are to deliver the performance improvements that end users are actually demanding, rather than the performance improvements the supply-side firms think or assume they are demanding.

Even *The Construction Performance Driver – A health check for your business* falls into this trap in its use of the term 'partners' because it fails to define precisely what is meant by the term. Does it refer to partnerships with end-user customers? Or does it refer to partnerships with other supply-side firms in the design and construction supply chain? In the conclusion of an excellent book by Mike Murray and David Langford (published by Blackwell Publishing) *Construction Reports 1944 – 98* and under the heading 'Recurring Themes' they make the point that *'the most famous buzzword of all, partnering, has been hijacked by consultants and corrupted by contractors'*.

To illustrate this confusion, in my discussions with industry firms (particularly the specialist suppliers and manufacturers that form the supply chain of the construction contractors), they have exposed a need to clarify what is meant by the term 'customer'. All too often the term 'customer' is incorrectly assumed to mean the next firm up the supply chain and it is assumed that only the needs of this supply-side customer have to be satisfied. In reality, the most important customer is the one that will occupy the building or use the facility – after all, there is little point expending money and resources on an improvement process that leaves the true end user as dissatisfied with the building or facility as they were before you started to improve your performance.

Similarly, what do end-user customers mean by the term 'value for money'? This is another buzzword that is freely banded about by supply-side firms, particularly in regard to an aspect of supply-side business performance that is important to demand-side customers. Yet I frequently hear supply-side firms claim that lowest capital cost is 'value for money' even though the term is very clearly defined in the *Charter Handbook*, *Modernising Construction* and *Better Public Buildings*. More worryingly, I even hear demand-side clients (more particularly the property divisions of demand-side clients) claim that lowest capital cost is 'value for money'.

Probably the best definition of 'value for money' is given under the 'Quality' heading in the Confederation of Construction Clients' *Charter Handbook* and is as follows:

☐ *'Aiming for quality-based solutions that yield maximum functionality for optimum whole-life cost, whilst preserving respect for the surroundings and the community.*
☐ *Promoting process and product improvements to minimise defects over the whole life of the construction solution.'*

The Confederation of Construction Clients' publication *Whole Life Costing – A client's guide* that came out about two years before the *Charter Handbook*, also gave an excellent definition of 'value for money':

'Major clients and leading industry figures have identified whole-life costing as the key to delivering improved value for money. It encourages the choice of a solution that strikes the most cost-effective balance between capital and running costs and minimises the risk of premature failure or loss of functionality of the construction on the business.'

This critically important need for supply-side firms to fully understand what is important to the demand-side end-user clients is picked up very clearly in Section 2 of *The Construction Performance Driver – A health check for your business*.

Under the sub-section heading 'Customer Results' it poses the following questions:

❑ 'Question 6.0 – Do you have data that demonstrates which features of your overall business performance are important to your customers?

❑ Question 6.1 – Does feedback obtained directly from customers indicate that they are satisfied with your products and services, the level of defects, cost and time predictability, and other agreed performance measures?

❑ Question 6.2 – Do you know if customer satisfaction levels are better than those of your competitors?'

The key to understanding the terms 'value for money' and 'customer' lies in the definition of quality from the *Charter Handbook* that I have given above. The two bullet points make absolutely clear that it is those that gain direct commercial benefit from maximum functionality and optimum whole-life cost and performance that are the true 'customers', and these true 'customers' are not surprisingly very concerned about the things that impact directly on the underlying labour costs and the overheads of their business.

Poor functionality (as the USA National Construction Goals emphasised) impacts very directly on the underlying labour costs and therefore the competitiveness of the end user. If the design of the building is such that the end user has to utilise 20% more people to do the work (the USA National Construction Goals report said that this could be as high as 30%), the price charged by the end user for its services or products will be far higher than was anticipated in the end user's business plan and will therefore damage the end user's profit margins and competitiveness.

Similarly, high levels of inefficiency and waste in the utilisation of labour and materials by supply-side firms in the design and construction supply chain will add to the initial capital cost and will therefore add to the end user's overheads, since the end user ultimately pays that additional cost (either directly or through rental charges).

In addition, defects in functional performance or defects in components or materials (or in the way they are put together in the construction process) over the intended life of the building will add directly to the end user's overheads, since it will inevitably fall to the end user to pick up the cost of the defects. As it is unlikely that the end user has budgeted for such defects, these will have to be met out of the profit margin or by raising the charge for the service or the price of the products produced by the end user. If the defect is big enough, it may even cause the bankruptcy of the end user.

This particular aspect of performance ought to be of great concern to any construction contractors involved with Public Finance Initiative (PFI) projects, although the following case history demonstrates that this is not always the case.

Case history – PFI hospital

An insurance company had been asked if they would insure a newly completed PFI hospital against premature failure of components or materials over the life of the PFI contract (25 years). Before quoting the annual premium, the insurance company insisted on carefully auditing the components and materials used and the way they had been put together in the construction process.

During the audit they discovered that during the construction process the construction contractor's buying department had been under pressure to reduce the initial capital cost because it had increased over the original budget figure. One of the strategies the buying department had adopted to save initial capital costs was to ignore the original specification for the fire dampers in the ventilation ducts. The original specification called for fire dampers that automatically reset themselves into the open position after they had been closed during a fire test. The buying department discovered that they could buy fire dampers for the ventilation ducting that had to be manually opened at a far lower price and that this cost saving would deliver the overall reduction in the initial capital cost they sought.

Unfortunately, the insurance company's audit exposed the fact that shortly after the hospital was opened and was in full operation a normal fire test was carried out. During the fire test the fire dampers in all the ventilation ducts were closed to prevent smoke from the fire spreading through the ducts to all parts of the hospital. When the fire test was complete, the facility management staff discovered that the only way the fire dampers could be opened was to take down the ceilings in the hospital corridors and reset the fire dampers in the open position.

Not only was this a time-consuming and expensive operation for the facility management staff (that would have to be repeated every time the fire test was carried out), it seriously hindered the effective operation of the hospital because the corridors had to be closed while the ceilings were taken down and fire dampers were opened. As the terms of the PFI contract demanded that the hospital accommodation had to be freely available 24 hours a day and 7 days a week, the PFI contractor had little option but to replace all the manual opening fire dampers with automatically opening fire dampers. The cost of this replacement was obviously a great deal more than the original saving in initial capital cost.

The lesson from this case history is that the construction contractor's buying department's failure to appreciate that keeping down the cost of ownership was far more important than keeping down the initial capital cost, had damaged the profitability of the PFI contract, had seriously damaged the operation of the hospital and had seriously damaged the credibility of the PFI approach to hospital construction.

When constructing an action plan for radically improved performance, supply-side firms should always remember that others have already trodden this road. It is my experience that those that have been through the pain barrier and survived are always delighted to share their experiences. The following case histories illustrate this willingness to share and the value of learning from others. They also illustrate that there is much that can be learned from other sectors of industry because the road is remarkably similar no matter what the nature of the business.

Case history – Aerospace industry

A major international firm from the aerospace industry achieved a dramatic improvement in performance in a remarkably short space of time.

The Chief Executive likened his role to that of a crusader king. He said he had to be constantly seen, by every one of his troops, to be leading the way forward into battle. Every move he made, every phrase he spoke and every word he wrote had to reinforce and clarify the changes in working practices he wanted the firm to make. He had to ensure that everyone in the firm (and he said this must literally include every last person employed in his firm) understood where the firm was going, why it must go there, what would happen if it failed to reach its destination, and (most important) what each individual had to do differently as their part of the change process. He said it was imperative that the tea lady and the cleaners felt they were included in the change process. They must understand why the changes in working practices were commercially essential and they must want to be an active part of the change process.

He also said that it was important to recognise and reward those that were making the greatest contribution to the change process. Quite often, the reward need be no more than public recognition by the Chief Executive for their efforts (i.e. a personal letter from the Chief Executive which is also put into the firm's newsletter).

He also said it was essential to provide everyone with regular progress reports which explained how the improvements in performance were being measured and what was being achieved in the various parts of the firm.

Equally important was the need to expose and deal with those that were blocking and opposing the changes in working practices. He said you could be absolutely certain that the grapevine would ensure that everyone would be aware of the names of those who were trying to block the changes. It was equally certain that everyone was watching to see if the Chief Executive was on the ball and would pick up on what was really happening. If the blockers were ignored, the message the grapevine would take around the firm was that the Chief Executive was not serious about the changes in working practices, so they could be safely ignored.

With regard to the following case history, and while I do not suggest that Defence Estate's drive to improve its performance is the ideal model for all to follow, it does provide some useful pointers on the things that initially inhibited the improvement process within Defence Estates and the things that had to be done to break down the barriers and facilitate progress. These facilitators and inhibitors are useful because I am assured that they are encountered time and time again by organisations in all sectors that are endeavouring to drive forward radical change.

Case history – Defence Estates

The Defence Estates drive for radical improvement began with the arrival of a new Chief Executive who had been drawn from the end-user side of the UK Ministry of Defence and had already been closely involved in a successful improvement drive in another part of the Ministry.

At his first Board of Directors meeting the Directors were explaining the effectiveness of the current procurement model in terms of lowest capital cost, achievement of time deadlines and the closeness of out-turn costs to tender prices (all of which compared very favourably with other major repeat clients). The new Chief Executive stopped the presentation partway through and made it very plain that he viewed the current procurement model as a total disaster.

He pointed out that he was from the end-user side of the Ministry and his experience (which he insisted matched the experiences of all other end users in the Ministry) was that the functionality of the buildings and facilities procured by Defence Estates was generally poor, that the whole-life costs were never predicted, that the on-site construction activities always appeared to be highly inefficient with long periods of inactivity or reworking, and that the operation of buildings and facilities was all too often beset with nasty surprises from the premature failure of materials and components.

He made explicit that the current approach to procurement had to be radically changed to give the end users what they really wanted, not what Defence Estates erroneously thought they wanted. He wanted a procurement approach that gave the end users the ability to buy buildings and facilities the way they bought all other products, i.e. a simple, one-stop approach to design and construction that enforced total integration of the design and construction process and of the design and construction supply chain, that delivered the maximum whole-life functionality for the lowest optimum whole-life cost, and that eliminated the inefficient utilisation of labour and materials (which of course precisely accords with the procurement approach recommended in *Rethinking Construction*).

He also made explicit that those Directors who were unwilling to commit wholeheartedly to his drive for improvement should leave Defence Estates.

The radical change programme that sprang from the arrival of the new Chief Executive was a long, hard journey with many setbacks and surprises. There were surprises in the unexpected barriers that held us back and in the things we had to do to eliminate those barriers. These inhibitors and the facilitators were as follows.

The inhibitors
- ❑ Covert blocking by senior managers, despite their public endorsement of the new way of doing things.
- ❑ Powerful inertia of traditional customs and practices.
- ❑ Deep-seated resistance to change at all levels.
- ❑ Inappropriateness of existing training that was geared to traditional customs and practices.
- ❑ Lack of any training that was directly supportive of the new procurement model.
- ❑ Ambiguously defined goals that were not understood by everyone in the organisation.
- ❑ Poor two-way communications within the organisation.
- ❑ Virtually no measurement of current and past performance.

❑ Unwillingness to listen to end users or suppliers.
❑ Lack of humility and a refusal to accept that anything was wrong with the way we had always done things.
❑ A belief that if you stonewalled long enough the pressure to change would go away.

The facilitators
❑ Committed and passionate leadership from the Chief Executive.
❑ The development of unambiguous goals for the improvement process.
❑ The introduction of accurate performance measurement using the 'award simulation' self-assessment system from the EFQM Excellence Model.
❑ The introduction of unfiltered feedback from end users and suppliers.
❑ The development of appropriate coaching and training for all staff using skilled and experienced external consultants.
❑ The development of good two-way communication (including its validation in use).
❑ The introduction of an honourable exit policy for those unable to change.
❑ A very public statement from the Chief Executive's boss that if Defence Estates refused to change it would be wound up.

The construction industry's traditional split between design and construction poses further problems for construction industry supply-side firms that no other sector has to overcome. Elsewhere design and manufacture are locked together within the supply-side supply chain, and the demand-side customer is never required to procure design separately from manufacture.

Unfortunately, in the fragmented and adversarial construction industry, those firms involved in design (architects, civil/structural engineers, mechanical/electrical engineers and quantity surveyors) generally see themselves as being an interface between the construction industry and the end-user client and not an intrinsic part of the construction industry. Consequently, any drive to improve performance and deliver buildings that meet the functional, whole-life, high labour efficiency, low materials waste aspirations of end users is unlikely to succeed without the entire supply-side design and construction supply chain agreeing to work together as an integrated supply-side partnership and also agreeing which firm is to lead and direct the supply-side partnership's drive for improvement.

This important (if not critical) issue of leadership of the integrated design and construction team needs to address the drive for improvement at two levels:

(1) At project level, which must include both the design and the construction stages.
(2) At strategic level, within and across the firms that make up the design and construction supply chain and from which the individuals that make up the project design and construction teams are selected.

Whilst I explained the key importance of leadership in an earlier chapter, the reality is that establishing the leadership function is fraught with difficulty and is beset with as much confusion over who should make the first move as all other aspects of improvement in the construction industry.

Many supply-side firms seem to believe that they should do nothing until they are forced into changing and improving their performance by major, repeat end-user clients and thus the drive for improvement should be led by the end-user client. Others seem to believe that the drive for improved performance only relates to construction activities and thus only concerns construction contractors and their supply chains of specialist suppliers (sub-contractors, specialist/ trades contractors and manufacturers) and should logically be led by construction contractors.

Many demand-side clients and supply-side firms seem to think that the drive for improvement can only be exercised within project partnerships between client and supply-side project teams, where the client representative in the team exercises the leadership role.

The result of this confusion is that everyone appears to be waiting for someone else to take the lead and very little gets done in the way of major and radical improvement. As a consequence, yet another report gets written to try and kick-start real change (the *Constructing the Team* report had little effect so the *Rethinking Construction* report was written; this in turn was followed by the *Accelerating Change* report – and this has been going on since the Simon Committee report in 1944).

The obvious reality of the continuous improvement process is that the drive for improvement will only be self-sustaining if it is supply-side led and independent of the end-user client. That way the improved performance will feed through to all constructed products for all end-user clients, the one-off and occasional end-user clients (including householders wanting an extension to their house) as well as the major repeat end-user clients. It also means that the improved performance will feed through to the maintenance and refurbishment of buildings and facilities as well as to the construction of new buildings and facilities. Again, this must feed down to the individual householder and not just apply to the major clients.

Continuously improving performance needs to be the way that the supply-side firms of all sizes (design firms as well as construction firms) do their business better for all their end-user clients, large and small, repeat as well as one-off and occasional.

As a consequence, the drive to improve performance will work best if it is a supply-side initiative that is led by a supply-side firm. But whilst it may be easier for construction contractors to assume the leadership role, it is not the only way that the drive for improvement can succeed and there is no reason why any member of the supply-side design and construction supply chain could not assume the leadership role.

In fact, since the leadership of the drive to improve performance will almost certainly be most effective if done within the 'Virtual Firm' where the design consultants, the specialist/trades contractors and the manufacturers are tied together in very long-term strategic supply-side partnerships, a question mark arises about the value that a construction contractor brings to the operation of the 'Virtual Firm'. After all, very few construction contractors directly employ the operatives that construct the building or facility and the reality is that about 95% of the resources required to design and construct the building or facility come from firms that are totally independent of the construction contractor.

In the long-term supply-side partnerships that form a 'Virtual Firm', the individual design and construction firms will quickly become mutually supportive and will become used to collaborating to co-ordinate their work at both strategic and project level. In these long-term supply-side partnerships, even project management could be drawn from within the various design consultancy or specialist supplier firms, as the particular project dictates.

As a consequence of this reality, perhaps the leadership for the drive for the improved performance of the design and construction supply chain would be better coming from that part of the design and construction team that provided it before the split between design and construction was made by the creation of the architectural profession. Perhaps the architectural profession ought to re-assume the leadership mantle of the construction team and accept that it naturally falls to them to assure the delivery of all the end user's aspirations, including the delivery of maximum functionality for the lowest optimum whole-life cost and the elimination of the unnecessary costs generated by the inefficient utilisation of labour and materials throughout the lifetime of the building (construction as well as maintenance activities).

One thing is certain: since the drive for improved performance must include the entire design and construction supply chain, and since it will work best if done through long-term strategic supply-side partnerships that involve every firm in the design and construction supply chain, one of the first questions that must be addressed by the construction contractors is what ought their role to be in the totally integrated supply-side team or 'Virtual Firm'? How can the construction contractor best add value to the overall performance of the design and construction supply chain? What would be the consequences, in terms of the overall performance of the design and construction supply chain, of the construction contractor not existing in the 'Virtual Firm' or in the design and construction supply chain?

After all, if the end-user client is selecting a totally integrated design and construction team from the outset of the design and construction process, and requires that team to produce evidence of improvements achieved within existing supply-side partnerships (i.e. is selecting a 'Virtual Firm' that has been together for some time), there is no obvious reason why the contact with the design and construction team should be via a construction contractor. In fact, the following case history from the Defence Estates' prime contracting initiative may be a marker for a possible way ahead.

> **Case history – Defence Estates' prime contracting initiative**
>
> The Defence Estates' prime contracting initiative was a one-stop shop approach to procuring the design, construction and maintenance of buildings that included the responsibility for the first five to seven years of maintenance after the completion of the building or facility.
>
> In one of the early prime contracts Defence Estates was approached during the 'expressions of interest' stage by a major quantity surveying consultancy that asked if they might put forward a response. They pointed out that the main requirement in the formation, direction and operation of

the design, construction and maintenance team was effective project management skills and they could demonstrate an excellent track record in just such an area of expertise.

The Defence Estates response was that they had never stipulated which firm in the design, construction and maintenance team should take the lead and form the contact point between the Ministry of Defence demand-side team and the industry's supply-side design and construction team. They would therefore examine each response on its merits and there was no reason why the major quantity surveying consultancy should not front a bid.

In private, Defence Estates were delighted that the initiative to lead the formation of a prime contract bid was being taken by a member of the supply-side other than a construction contractor. The hope was that when the industry learned of the bid by the major quantity surveying consultancy, it would go a long way towards shaking construction contractors out of their apparent complacency.

When all the expressions of interest were carefully and objectively evaluated and marked, the submission from the major quantity surveying consultancy was selected to go forward to the next stage.

When the final stage of the selection process was reached, the bid from the major quantity surveying consultancy was a clear winner. They had put together an excellent design, construction and maintenance team. They understood what improvements in functional and whole-life performance and cost Defence Estates and the end users demanded, and they put forward a coherent strategy for delivering those improvements. Their submission clearly understood that what was sought by Defence Estates was a response from a pre-formed team that included and named all the key supply-side firms. Their submission understood that such a pre-formed team was completely different from the usual situation where a construction contractor puts forward a bid in virtual isolation from the specialist suppliers that will design and construct the building, and in total isolation from the firms that will maintain the building after completion.

SUPPLY-SIDE ACTION PLAN FOR THE INTRODUCTION OF PERFORMANCE MEASUREMENT AT STRATEGIC LEVEL

Once a leading member of the supply-side design and construction supply chain has decided to take the lead in embracing performance measurement as a means of driving forward radical improvement, and once that firm has decided to start the improvement process by utilising the project level guidance given in Chapter 7, Performance Measurement at Project Level, on one or two test-bed projects, the next move is to develop an action plan that enables the export of the experiences and the lessons learned from the test-bed projects to all other projects undertaken by the design and construction supply-side 'Virtual Firm'.

The action plan will need to address the necessity of changing the way all the firms and all the personalities in the design and construction supply chain work together. The action plan needs to enforce a far greater degree of co-operation and collaboration within project teams and between supply-side firms than is usually the case. A possible mechanism for achieving this was dealt with in Chapter 5 on the 'Virtual Firm' and the formation of a 'Virtual Firm' ought to be the first priority of the action plan and the improvement process.

Strategic Level Action Plan Stage 1 – Laying the foundations

The first stage of the Action Plan must be the validation of the meaning of all the various terms (such as CSF and KPI), the validation of the destination or the goals of the action planning process and the validation of the foundations of the improvement process. There is very little point in spending a great deal of time, money and resources changing the way you and your supply chain operate if the end result still fails to satisfy the demands of the end users. The responses to the following questions may help the validation process:

- ❏ Are you absolutely clear about what the intended drive to improve performance is intended to achieve and why you need to achieve it?
- ❏ Do you know how your drive to improve performance will impact on the firms in your supply chain?
- ❏ Do you intend that your drive to improve performance will be done independently of your demand-side end-user clients? If not, do you have major, repeat end-user clients who are enthusiastic about helping the supply-side to improve its performance and with whom you could form strategic partnerships?
- ❏ If you are relying on strategic partnering with major, repeat end-user clients to provide the impetus for your drive to improve performance, how do you intend to transfer the performance improvements to other end-user clients, especially the one-off end-user clients?
- ❏ Are you prepared to fully accept and openly admit to serious shortcomings in performance in your own firm and that of your supply chain firms, such as inefficient utilisation of labour and materials, poor whole-life costs, poor whole-life performance, poor durability and poor functionality?
- ❏ Are you and your supply chain firms clear about what is meant by the key terms used in your drive to improve performance?
- ❏ Can you validate the meaning of all the key terms used in your drive to improve performance, against definitions taken from publications that are universally acknowledged by the end users of constructed products to fully reflect their commercial needs?
- ❏ Do you have a mechanism in place to select preferred supply chain firms from your full database of firms and set up strategic supply-side partnerships?

❏ Have you the resources to work with, and actively include, all your preferred supply chain firms in your drive for improvement?

❏ Do you and your supply chain firms have a standardised method of measuring the effective utilisation of labour and materials, the satisfaction of end users with functionality, and the post-occupation/completion defects?

❏ Have you aligned your improvement goals with the improvement goals demanded in key end-user sponsored publications, such as the *Charter Handbook* or *Modernising Construction* (or even the USA National Construction Goals if your firm operates in the USA)?

❏ Does every Chief Executive of every supply-side firm in your design and construction supply chain share a common understanding of what aspects of performance need to be improved?

❏ Is every Chief Executive of every supply-side firm in your design and construction supply chain as equally committed as you to the achievement of your improvement goals in the timescale you have set?

❏ Do you have (and have you tested and validated) a communications mechanism in place throughout your own firm, and the firms of your supply-side partners, that ensures everyone shares a common understanding of the goals in your drive for improvement, of the reformed processes by which those goals are to be achieved, of the means by which improved performance will be measured, and of the commercial justification for setting the goals?

❏ Can you produce evidence to prove that you are truly 'walking the talk' and not just 'talking the talk'?

❏ Do you have a mechanism in place for assuring that all your senior managers are on side and are as equally committed as you to the attainment of the goals in your drive for improvement?

❑   Do you have a mechanism in place that enables you (as Chief Executive of the supply-side firm that is leading the drive for improved performance) to constantly check what is really happening at the sharp end at project or site level?

Strategic Level Action Plan Stage 2 – Forming the Virtual Firm

Actions from my book *Building Down Barriers – A guide to construction best practice* describe the process by which a virtual supply-side firm could be formed and are listed on p 112.

Strategic Level Action Plan Stage 3 – Introducing Performance Measurement at Project Level

Having established the foundations of the improvement process and started the process of forming a 'Virtual Firm', the next move for those firms within the embryonic 'Virtual Firm' is to set up a mechanism for measuring performance and driving forward improvement at project level across all the design and construction firms within the embryonic 'Virtual Firm'. It is important to remember that to be effective this must also include those firms in the supply chains of the first-tier firms, i.e. in the case of the mechanical and electrical services firm, the performance measurement process must be adopted by every firm that supplies services, materials and components to the mechanical and electrical services firm.

Radical performance improvement that is intended to drive out unnecessary costs, drive up whole-life quality and drive down whole-life costs will only work if it is done by every firm in the supply-side design and construction supply chain, so that all have the same goals in mind and are using the same performance measurement system. This obviously requires the performance measurement system to be an intrinsic part of the strategic supply-side partnerships that form the basis of the formation of the 'Virtual Firm'. It also obviously requires all the design and construction firms to agree common aims and objectives to ensure that all are going in the same direction at the same time.

In order to do this, they must all agree what aspects of their own and each other's performance are in most need of improvement, which of course means they all have to agree which aspects of their own and each other's performance are least effective and what the consequences of this less-than-effective performance are in terms of the improved output deliverables now demanded by end-user clients (i.e. the elimination of unnecessary costs and the delivery of the best whole-life quality for the lowest whole-life cost).

A detailed action plan for the introduction of performance measurement at project level was dealt with in Chapter 7.

Strategic Level Action Plan Stage 4 – Introducing Performance Measurement across Supply Chain Firms

Each and every supply-side firm in the 'Virtual Firm' ought to be continuously measuring their performance and looking for ways of improving that performance at every level, from individual project level up to the business activities within the firm that over-arch and support the individual projects. This will obviously necessitate those supply-side firms that have their own supply chains working with the firms in their supply chains to continuously measure each firm's effective utilisation of labour and materials, and to continuously benchmark (compare) each firm's performance (especially in terms of process and technical innovations) with their competitors.

As I have mentioned before, the UK Construction Best Practice Programme has worked with the British Quality Foundation (BQF) Construction Group and BQC Performance Management Ltd to produce *The Construction Performance Driver – A health check for your business*. This excellent, simple-to-understand and concise guide was written using feedback from the BQF Construction Group members who are at the sharp end of change management and performance improvement within construction industry firms.

There is little point in my reinventing such an excellent wheel, so I strongly recommend that *The Construction Performance Driver – A health check for your business* should be used as the basis of this stage of the action plan. However, I suggest the following questions from the various sub-sections should be given particular attention by supply-side firms within the 'Virtual Firm':

Section 1 – 'The Enablers' – 'Policy and Strategy' sub-section:

❑ *'2.3 – Do you review the activities and successes of your major competitors when you make or review your plans?*

❑ *2.4 – Do you look at the way that 'best companies' do the things you are trying to do?*

❑ *2.12 – Do you make sure that you are aware of any new technological developments which may affect your business?*

❑ *2.13 – Have you identified all of the elements of your operation (processes) that are key to your success?'*

'The Enablers' – 'Partnerships and Resources' sub-section:

❑ *'4.6 – Do you work with key partners and suppliers to help them improve their performance?*

❑ *4.7 – Do partners and suppliers consistently deliver what is required to meet your needs?*

❑ *4.9 – Is waste of all kinds minimised (e.g. in both material and business process waste)?'*

'The Enablers' – 'Processes' sub-section:

❑ *'5.5 – Do you measure how well your processes work?*

❑ *5.6 – Are processes continually being improved?*

❑ *5.7 – When problems occur, do you look for the real cause so that you can implement innovating long-term solutions rather than quick fixes?*

❏ *5.9 – When you make changes to the way you do things, do you review how well the changes have worked?'*

The final message in this chapter has to be the self-evident message that the above action plan will not happen by magic. It will only happen if one of the supply-side firms that makes up the design and construction supply chain actively takes up the leadership mantle and makes the action plan happen. It must also be borne in mind by every firm on the supply-side that no supply-side firm can assume a right to lead the drive for improved performance in a situation where that improved performance must come from the entire supply-side design and construction supply chain. The leadership mantle is available for any firm to take up and the all-important need is to ensure that all the firms on the supply side agree to co-operate enthusiastically and willingly with the firm that wishes to take up the leadership mantle.

The leadership role will work best if it is earned, not if it is imposed. It must be earned by demonstrating to the other supply-side firms a deep and comprehensive understanding of what needs to be improved across the entire design and construction process, why it needs to be improved, who needs to be involved in the improvement process, how the improved performance can be sustained and delivered for all end-user customers, what the end result of the improvement process needs to deliver in terms of end-user aspirations, and how the improved performance will be measured.

Above all, the firm that proposes to adopt the leadership mantle of the improvement process ought to be able to demonstrate the deepest understanding of the Rethinking Construction six primary themes of construction best practice:

❏ *'The finished building will deliver maximum functionality and delight the end-user.*
❏ *End-users will benefit from the lowest optimum cost of ownership.*

❑ *Inefficiency and waste in the use of labour and materials will be eliminated.*
❑ *Specialist suppliers will be involved from the outset to ensure integration and buildability.*
❑ *Design and construction will be through a single point of contact.*
❑ *Performance improvement will be targeted and measurement processes put in place.'*

9 The Client's Selection Process

The end-user client's process for the selection of the supply-side design and construction team is a key factor in driving forward radical change. For as long as the demand-side client appoints the design team separately from the construction team and uses lowest price as the basis of selection, the industry cannot fully integrate the design and construction team, those firms delivering the design and construction services will be under pressure to cut corners, and the construction contractor will see no reason why lowest price selection should not be used for the appointment of the specialist suppliers (sub-contractors, trades contractors and manufacturers).

In fact, it is highly likely that unless the end-user clients change their procurement practices to a form that encourages and facilitates integration of the design and construction supply chain and the delivery of the highest optimum whole-life performance for the lowest optimum whole-life cost, the current drive for improvement will quickly die away.

In earlier chapters I reminded you that in the *Accelerating Change* report, Sir John Egan had stated:

'Clients should require the use of integrated teams and long term supply chains and actively participate in their creation.'

I also reminded you of the Rethinking Construction guidelines listed in their publication *Rethinking the Construction
Client – Guidelines for construction clients in the public
sector*. The six guidelines define the targets that public
sector clients seeking best value should aim for when procuring constructed products such as new buildings, refurbishment, adaptations and maintenance. I pointed out that the
guidelines are equally valid for private sector clients that are
determined to achieve the best whole-life value for the
lowest whole-life cost. The guidelines are given as:

❑ *'Traditional processes of selection should be radically
changed because they do not lead to best value.*
❑ *An integrated team, which includes the client, should be
formed before design and maintained throughout delivery.*
❑ *Contracts should lead to mutual benefit for all parties and
be based on a target and whole-life cost approach.*
❑ *Suppliers should be selected by Best Value and not by
lowest price: this can be achieved within EC and central
government procurement guidelines.*
❑ *Performance measurement should be used to underpin
continuous improvement within a collaborative working
process.*
❑ *Culture and processes should be changed so that collaborative rather than confrontational working is achieved.'*

Earlier chapters have explained the consequences for the
UK construction industry of the publication of the Confederation of Construction Clients *Charter Handbook*, the National Audit Office report *Modernising Construction* and
the Department of Culture, Media and Sport report *Better
Public Buildings*.

The existence of these three publications, particularly
Modernising Construction and *Better Public Buildings*,
means that all public sector clients (who represent 40% of
the total UK construction market) will inevitably have little
option but to adopt the best practice approach described in
the National Audit Office and Department of Culture, Media

and Sport reports. They are all subject to external audit by the National Audit Office (for central government procurers of construction services) or by the Audit Commission (health authority and local government procurers of construction services) and as these two audit bodies will inevitably use *Modernising Construction* and *Better Public Buildings* as the common evaluative criteria for all their assessments, those audited by them will have little option but to adopt the same approach to best practice.

The private sector will obviously have far greater freedom to choose an appropriate approach to procurement. However, if they choose to adopt the principles of the *Charter Handbook*, as the *Accelerating Change* report strongly advocates, their procurement practices will need to be reviewed against the six goals of construction best practice and changed where they diverge.

In earlier chapters it was made clear that the *Charter Handbook*, *Modernising Construction* and *Better Public Buildings* have similarities and throw up the same two key differentiators and the same six primary goals of construction best practice. These are:

The two key differentiators of construction best practice

☐ *Abandonment of lowest capital cost as the value comparator.* This is replaced in the selection process with whole-life cost and functional performance as the value for money comparators. This means industry must predict, deliver and be measured by its ability to deliver maximum durability and functionality (which includes delighted end users).

☐ *Involving specialist contractors and suppliers in design from the outset.* This means abandoning all forms of traditional procurement which delay the appointment of the specialist suppliers (sub-contractors, specialist contractors and manufacturers) until the design is well advanced (most of the buildability problems on site are created in the first 20% of the design process). Traditional forms of sequential appointment are replaced with a requirement to appoint a totally integrated design and construction supply chain from the outset. This is only possible if the appointment of the integrated supply chain is through a single point of contact – precisely as it would be in the purchase of every other product from every other sector.

The six primary goals of construction best practice

☐ *'The finished building will ensure maximum functionality.*
☐ *The end users will benefit from the lowest cost of ownership.*
☐ *Inefficiency and waste in the utilisation of labour and materials will be eliminated.*
☐ *The specialist suppliers will be involved in design from the outset to achieve integration and buildability.*
☐ *The design and the construction of the building will be achieved through a single point of contact for the most effective co-ordination and clarity of responsibility.*
☐ *Current performance and improvement achievements will be established by measurement.'*

In Chapter 6 the actions required to motivate and lead radical change within any organisation and within any sector of the construction industry were explained and it was emphasised that it was of paramount importance for the Chief Executive to adopt a powerful crusading role if the change was to succeed. The Chief Executive of the demand-side client organisation must ensure the following four ingredients are in place:

☐ A clearly explained and rational goal for the reformed procurement process that all can understand, with which all can identify, and which can be related to specific improvements in performance that can be measured and compared with current performance.

- ❏ Committed, determined and overt leadership by the Chief Executive, which leaves no-one in any doubt about where the reformed procurement process must take the organisation, why it is essential to go there, and the timescale for the change.
- ❏ A detailed and comprehensive action plan for the development and implementation of the changes in working practices, which explains in simple, easy-to-understand language what must be done differently by every member of the organisation. Adequate and appropriate training must support this for those that are required to operate in a different manner.
- ❏ A simple and easy-to-understand explanation of the benefits that will be delivered to the end users by the changes in working practices. This is best expressed in terms which relate to the aspects that must be improved, namely the elimination of unnecessary costs in the construction team's utilisation of labour and materials and the delivery of the highest whole-life quality for the lowest optimum whole-life cost.

The following case histories provide considerable evidence that the adoption of the supply chain management principles that are common to the *Charter Handbook*, and to the *Modernising Construction* and *Better Public Buildings* reports, will deliver major commercial improvements to both the demand and the supply side of the construction industry. Other major repeat clients in the UK, such as BAA, McDonald's and Argent, have actively taken control of their supply chains and used supply chain management techniques to deliver major commercial improvements for themselves and their supply-side partners. These include reductions in the capital cost, better durability, fewer post-completion defects and less reworking.

Not only does the adoption of supply chain management principles deliver improved lifetime quality and functionality and reduced capital and operating costs to the end user, but the profits made by the supply-side firms will be enhanced by converting inefficiency and waste in the utilisation of labour and materials into lower costs and higher profits.

These case histories provide hard evidence of the considerably better value that can be achieved if the traditional lowest price tendering is replaced by a selection process that focuses wholly or primarily on the experience, the knowledge and the skills of a fully integrated supply chain, especially those of the specialist suppliers (sub-contractors, trades contractors and manufacturers) and ensures (preferably enforces) the involvement of the specialist suppliers in design from the outset of the development of the design. The two case histories also show the magnitude of the improvements that can be delivered if the selection process also demands hard evidence to back up all the claims made about improved performance by the competing design and construction teams.

The St Helens Metropolitan Borough Council approach to value-based selection came out of recognition that lowest price tendering regularly failed to deliver best value. Out-turn prices regularly exceeded tender prices by a significant amount, buildings were regularly delivered late, maintenance costs were never predicted and regularly came out higher than expected, components and materials regularly failed earlier than expected, and end users were regularly dissatisfied with the functional performance of the completed buildings.

Since lowest price tendering regularly failed to deliver best value, St Helens (officers, elected members and internal audit staff) took the very courageous decision to abandon lowest price tendering. They decided that best value was most likely to be delivered if their selection process selected the supply-side team that could demonstrate, and prove through hard evidence, that it had the highest level of appropriate skills, experience and knowledge.

The selection process would announce and set aside from the selection process the project's gross maximum price (which generally came from externally imposed government cost limits). The selection process would then focus wholly on the skill and experience of the supply-side team by demanding evidence of the use of an open-book approach down through the supply chain, of effective supply chain management, of total integration of the design and construction supply chain, of effective application of value management and value engineering techniques, and of the involvement of end users in the value management and value engineering workshops. The selection process would also predetermine the mechanism by which savings would be shared between the end-user client and the supply-side team on a 50–50 basis.

The St Helens case history proves how effective their best value selection process was in the first project. There have now been many other projects based on the best value selection process, including several housing association projects, and all have demonstrated beyond any possible doubt that the approach out-performs all lowest price tendering processes.

In fact, the primary school in the case history was compared with another primary school close by, where the contract had been let shortly before the case history school and had been awarded on the basis of the usual lowest price tendering approach. Both were identical in size and identical in functional requirement. The out-turn price (final account) of the other primary school exceeded the tender price by around 11%. In addition, the tender price of the other school exceeded the gross maximum price of the case history school by around 9%, but the gap between the two schools was further widened because the best value approach delivered the case history school at around 18% less than the gross maximum price.

This meant that the case history school, using the best value selection process, was around 38% better value than the school that used lowest price tendering. In fact, the case history school was delivered with far better functionality than the other school because the supply-side team and the headmaster of the school (with St Helens Metropolitan Borough Council agreement) agreed that the 18% would be ploughed back into the school to give even better value.

Case history – St Helens Metropolitan Borough Council
The Borough Council introduced a 'Best Value' approach to construction tendering as an alternative to the traditional confrontational approach of the industry with its 'sealed envelope' lowest price process which failed to provide clients with value for money because projects would regularly exceed the original cost and completion time predictions.

Their 'best value' tendering approach appointed the entire design and construction supply chain from the outset, had a selection process that was based solely on the skills and experience of the entire design and construction team (including the specialist suppliers) rather than the traditional 'lowest price' approach, and included sharing savings with the supply-side team on a 50–50 basis.

When tested on a £2.3 million primary school, the out-turn cost to the Borough Council remained the same as the original cost estimate but enhanced standards amounting to £400 000 were able to be built into the project.

The school was completed ahead of the originally planned completion date, much higher mechanical and electrical standards than the norm were achieved and the site had an excellent safety record. The headteacher and his staff were involved in the partnering process from day one and were convinced that functionality was well above that usually delivered by new buildings.

Case history – Building Down Barriers pilot projects

The two pilot projects used as test-beds for the Building Down Barriers supply chain management toolset produced exceptional improvements in performance over conventionally procured buildings.

Labour efficiency on site reached 70%, which was almost double that usually achieved, and both specialist supplier teams came close to achieving a 'right first time' culture. Wastage of new materials was close to zero, whereas the industry norm is around 30%.

Both buildings were completed ahead of the contract completion date, with one being completed 20% ahead. The through-life cost forecast for both buildings was well below that for traditionally procured buildings, without incurring a significantly increased capital cost.

Functionality was demonstrated at every stage of design using 3-D visualisation, which avoided the usual problem of end users struggling to understand 2-dimensional drawings, and the end users were delighted with actual functionality of the completed buildings, which exceeded anything they had experienced before.

The St Helens Metropolitan Borough Council best value selection process was a two-part process that used carefully structured questionnaires marked by an in-house team to avoid any possible risk of accusations of subjectivity.

The stage one questionnaire asked a range of questions that included the following:

❑ Can you demonstrate partnering experience with other members of your supply chain?
❑ Do you benchmark against Construction Best Practice Programme Key Performance Indicators or other internal Key Performance Indicators?

❑ Do you benchmark your performance through post-contract (post-handover) questionnaires including client/customer satisfaction feedback?

❑ Do you encourage training through the Construction Skills Certification Scheme or other recognised schemes?

❑ Can you provide any evidence of your firm's commitment to change under Rethinking Construction?

❑ Can you demonstrate areas of innovation and good practice introduced by your firm?

❑ Can you demonstrate how your firm is dealing with the training necessary within your organisation to deal with the culture change required?

❑ List five key skills that you would expect your own and your supply chain's key personnel to possess as a group.

❑ What increased productivity do you envisage will ensue from good partnering practices?

❑ List up to six mechanisms that you use to ensure effective management of time-scales within projects.

❑ Give examples of how you identify best practice and how the information is used in your process of continual improvement.

❑ Explain what 'Respect for People' means in your organisation; outline its application and benefits to you and your supply chain partners.

The St Helens Metropolitan Borough Council marking system reduced the responses to the stage one questionnaire down to five or six firms who were then asked to complete a project specific questionnaire that sought method statements to set out how the supply-side team would approach the specific project.

This included the following:

❑ **The competency, understanding and skills of the supply-side team.** These were required to relate to the areas of expertise highlighted by the stage one question-naire.

- ❏ **Ability to change.** This related to the mechanism by which the supply-side team would apply well-established improvement processes (which included changing the culture of the supply-side firms) to improving the culture of any local firms that were introduced into the supply-side team to meet the Borough Council requirement to make maximum use of local labour.
- ❏ **Value management/value engineering.** This required the method statement for the application of value management techniques to define the precise end-user functional requirements, and the method statement for the application of value engineering techniques to drive out unnecessary costs in the use of labour and materials and to drive up whole-life quality.
- ❏ **Systems management.** This required the method statement for the management of all the design and construction processes and activities so that 'right first time' would be achieved on site.
- ❏ **Best practice and innovation.** This required the method statement that showed how best practice and innovation would be imported into the design and construction process for the specific project. This was especially important where new firms were introduced into the supply-side team to meet the Borough Council requirement to make maximum use of local labour.
- ❏ **Pre/post-contract costs.** This required the method statement that set out the management system that would control the underlying labour and material costs throughout the design and construction supply chain, especially the capturing and sharing of savings and the management of risks.

Both case histories emphasised the importance to the demand-side client of delivering the precise functional performance required by the end user, of using value management techniques to accurately capture and record the end user's functional requirements, and of involving the end user in value engineering workshops. The importance of delivering a level of functional performance that delighted the end-user was also a key theme of the *Charter Handbook*.

As a consequence, the demand-side client's selection process must (as the St Helens Metropolitan Borough Council selection process does) demand evidence from the supply-side team of the method they have used on previous projects to deliver maximum functionality. The method statement must include the value management techniques that were used to capture and record the end user's specific functional requirements, the value engineering techniques that ensured that the functional requirements were fully incorporated into the developing design, and the measurement system that was used to test the functional effectiveness of the completed building.

The importance to the end users of the efficient functional performance of the completed building is not unique to the UK. The same concern can be found in virtually all developed countries and was expressed very clearly in the report in 1996 that set the USA National Construction Goals. The USA report made the point that if the salaries of the occupants are included in the running costs, the running costs of a typical office block for a single year equal the capital cost of construction.

In fact, the USA National Construction Goals included a target for better design to deliver a 30% improvement in occupant efficiency. Since it is unlikely that design in the UK is delivering significantly higher occupant efficiency levels than in the USA, a similar improvement in UK occupant efficiency ought to be possible if the integrated approach described in all three standards (*Charter Handbook, Modernising Construction* and *Better Public Buildings*) is adopted by the UK construction industry.

The focus on delivering occupant efficiency is reflected strongly in the two key differentiators and the six primary goals of best practice listed at the beginning of this chapter. It does not require a great deal of logical thought to recognise the commercial benefit to the end users of buildings which enable the occupants to boost their efficiency by 30%, especially where the functional excellence and the environment are such that the morale of the end users is raised to a level where they are delighted to be working in or using the building. The *Better Public Buildings* report makes the very telling point that if hospitals provided a functionally excellent environment, patients would recover more quickly and would therefore be discharged earlier and this would represent a major operational saving for the hospital.

Both case histories also demonstrated that the use of performance measurement techniques within a continuous improvement regime can dramatically increase the effective utilisation of labour and materials and the evidence from the two case histories validates the claim made in the 1994 Latham Report *Constructing the Team* that inefficiency and waste in the utilisation of labour and materials consume 30% of the initial capital cost of construction.

Since it is highly unlikely that inefficiency in the UK is restricted to new construction activities, it is almost certain that construction activities on maintenance and refurbishment in the UK are also suffering a level of unnecessary cost due to the inefficient utilisation of labour and materials that amounts to 30% of the initial capital cost.

It follows from the above that there is an urgent and overwhelming need for a radical change in the sequential, fragmented procurement practice of clients and end users. All three standards (*Charter Handbook*, *Modernising Construction* and *Better Public Buildings*) see the consequences of sequential procurement as the primary cause of poor value from the construction industry.

The industry itself is not guiltless in the use of lowest capital cost and lowest professional fees as the basis of its selection process. Construction contractors regularly use lowest capital cost as the mechanism by which they select their specialist suppliers (sub-contractors, trades contractors and manufacturers) and design–build contractors regularly use lowest professional fees as the mechanism by which they select their design professionals.

Regardless of the sins of the construction industry in its addiction to lowest price selection of its suppliers, the demand-side clients need to follow the courageous example of clients like St Helens Metropolitan Borough Council and take the lead by replacing their lowest price tendering systems with value-based selection that focuses on skills rather than price. The pace of reform of the construction industry in the UK would be dramatically increased if far more demand-side clients followed the St Helens lead and utilised a similar value-based selection system.

In the case of public sector clients, the St Helens Metropolitan Borough Council value-based selection system has been proven beyond doubt and can be copied by any public sector client. In fact, since St Helens is now a Beacon local authority, under the Government scheme, they have a duty to help other local authorities copy their value-based selection system.

For those clients that wish to develop their own value-based selection system, there is still a need to adopt a degree of commonality with other clients to avoid the consequences of a confused free-for-all, where every client adopts a different selection system and the construction industry has to bear the cost of having to interpret, and respond differently to, each client's differing selection system.

As Rethinking Construction has given a lead to clients by adopting and promulgating the six goals of construction best practice that were first listed in the Construction Best Practice Programme booklet *A Guide to Best Practice in Construction Procurement*, it would make sense if demand-side clients also adopted the six goals and made them the basis of their selection system.

Thus demand-side clients that wish to abandon lowest price tendering and are determined to introduce a value-based selection system, should start the change process by willingly and enthusiastically embracing the two differentiators and the six goals of best practice procurement that are the foundation of the three standards.

This also applies to those professional advisors that provide an interface between the end users and the construction industry, whether employed internally by the client within a property division or employed externally as consultants, since they normally provide the expert advice on procurement methods and, quite often, they place the design and construction contracts with the industry. Quite often, it is the client's or end user's professional advisors that are responsible for the sequential appointment of the consultant designers and the construction contractors. Consequently, it is their actions and their advice to the client or end user that make it impossible for the specialist suppliers to have any input into the initial concept design, even though it is well known that most of the buildability problems on site are created in the first 20% of the design process.

The two key differentiators and the six primary goals of best practice enable those clients (and their professional advisors) who wish to embrace the principles of the Confederation of Construction Clients *Charter Handbook* to objectively review their current procurement practices to assess how closely they accord with the best practice of the three standards. Where any aspect of current practice is found to be at variance with best practice, action will need to be taken to improve the defective procurement practices.

One aspect of traditional procurement practice of some clients that directly inhibits supply chain integration is the practice of treating design as an in-house function that can safely be separated from construction. Where this occurs it is almost impossible for the specialist suppliers to be involved in design from the outset and thus it is almost impossible to harness their experience and skill to eliminate inefficiency and waste in the utilisation of labour and materials and thus achieve a 'right first time' culture on site. As a consequence, where in-house design continues to be retained by the client it is very unlikely that the client will ever achieve best value in construction procurement.

The manner by which the best practice of the three standards is introduced by clients will be different depending on the nature of the client's construction programme. The majority of clients are small and occasional procurers of construction and are not therefore in a position to become a major driver of the change process. The clients that are repeat procurers of construction have considerable leverage and can use their changed procurement practice to force the pace and direction of the change process. Excellent examples of this in the UK are BAA with their framework contract approach, Defence Estates with prime contracting, and NHS Estates with their Principle Supply Chain Manager approach. All three involve the introduction of supply chain integration and management by means that are appropriate to each client and all three believe that involving specialist suppliers in design from the outset and thus harnessing their considerable skill and experience while the design solution is still at concept stage, will deliver better value.

The following section briefly describes a possible change process that could be adopted by repeat clients and by small and occasional clients that are determined to abandon lowest price tendering and embrace value-based selection. It should be read bearing in mind the St Helens Metropolitan Borough Council and the Building Down Barriers pilot project case histories described earlier in this chapter. It should also be read bearing in mind the St Helens Metropolitan Borough Council value-based selection system that was described in some detail immediately following the case histories. The suggested change plan is as follows.

INTERNAL CHANGE PROCESS FOR DEMAND-SIDE CLIENTS WHO WANT TO EMBRACE VALUE-BASED SELECTION

An action plan to improve the demand-side client's procurement process was first described in my book *Building Down Barriers: a guide to construction best practice*. This has now been refined using feedback to improve its user friendliness.

Repeat clients

Assess whether you want the cost benefits of the construction best practice of the six primary goals, such as considerably lower capital and operating costs. These come from harnessing the very considerable knowledge and experience of the specialist suppliers from the outset of design development in order to eliminate inefficiency and waste in labour and materials utilisation, and from the use of high quality and durable materials and components. Construction best practice should provide accurately predicted, significantly lower and risk-free running costs that can confidently be built into long-term business plans. Construction best practice also delivers reduced operating costs that come from the impact that excellent functionality has on the efficiency and morale of the workforce. Construction best practice delivers greater cost and time certainty that comes from the actual out-turn cost never exceeding the forecast and the building always being completed on time.

To assess how well your current design and construction process accords with best practice, check if your construction contractors measure the efficient utilisation of labour and materials throughout their site activities. Speak to specialist suppliers (sub-contractors and trades contractors) about the frequency and causation of disruption to their work on site. Undertake end user surveys to test the effective functionality of completed buildings. Compare out-turn costs and actual completion dates with the initial budget and the forecast completion dates of a selection of completed buildings. Seek advice from your facility management staff on the maintainability, durability and ease of operation of completed buildings and facilities.

If the results of this assessment demonstrate a gap between your current design and construction process, and the best practice of the six goals, you would be wise to adopt the following action plan:

❑ Compare current practices and processes with the two key differentiators and the six primary goals of the three best practice standards i.e. does the current procurement process ensure the involvement of specialist suppliers from the outset of the design process, which is critical to a 'right first time' culture on site? If current practice starts with the appointment of an architect to develop the design, who then appoints a construction contractor, who in turn appoints specialist suppliers at a time when it is far too late for them to have any influence on the design (especially the elimination of inefficiency and waste in the utilisation of labour and materials), it is unlikely that the six primary goals will be delivered and you are inevitably paying a heavy cost penalty. Always ensure that this comparison is done objectively and never make assumptions or accept anyone's opinion as fact. It may be best to seek the assistance of an independent expert, who can demonstrate a comprehensive understanding of the six goals, to assist the comparative analysis.

❑ Once the comparative analysis has objectively established any variance between current practices and best practice, the next step is to set an improvement target for the organisation and to ensure everyone fully understands what the target means, why the target is important to the organisation, what benefits the improvements will deliver, how the improvements will be measured, and how the target will affect his or her individual role.

❏ Set Key Performance Indicators that enable you to accurately measure the organisation's rate of improvement, such as: end-user satisfaction with functionality; the production of, and the accuracy of, the cost of ownership predictions made by the design and construction teams; the rate of improvement in the on-site utilisation of labour and materials; the stage when specialist suppliers actually become involved in design development (especially whether their skill, knowledge and experience is really being used to the greatest advantage by the designers); and the speed of introduction of single point procurement.

❏ Ensure all leaders, especially the head of the organisation, become knowledgeable crusaders and champions for best practice. This requires them to have a deep and consistent understanding of the six primary goals and to ensure that every word they utter, every action they take, and everything they write, reinforces the change process.

❏ Communicate the improvement targets and the intended changes in current practices and processes to everyone within your own organisation and in those organisations with which you interface in the construction industry. This ensures that everyone fully understands where you are going, why you are going there and how you intend to get there. This is particularly important for those with whom you interface, because they need to work out how it affects their own practices and processes. It is imperative that the language used is such that everyone can understand the message; there must be no ambiguities and there must be some way of checking the understanding of the recipients. The responsibility is always with the sender of the message to use the most appropriate language and the KISS (Keep It Simple, Stupid) principle should always apply.

❑ Only invite expressions of interest from those design and construction firms that are able to provide well-documented evidence that they have practices and processes in place that ensure delivery of the six primary goals. In particular, demand the names of those firms (including the design and other consultants, but particularly including specialist suppliers) that can prove that they are already bound together within an integrated team in long-term strategic supply chain partnerships.

❑ Focus your in-house resources on defining the business need in terms that are suitable for measuring and evaluating the output performance of the built solution; leave designing the solution to the integrated design and construction team.

Occasional clients (large and small)

Assess whether you want the cost benefits of the construction best practice of the six goals, such as considerably lower capital and operating costs. These come from harnessing the very considerable knowledge and experience of the specialist suppliers from the outset of design development in order to eliminate inefficiency and waste in labour and materials utilisation, and from the use of high quality and durable materials and components. Construction best practice should provide accurately predicted, significantly lower and risk-free running costs that can confidently be built into long-term business plans. Construction best practice also delivers reduced operating costs that come from the impact that excellent functionality has on the efficiency and morale of the workforce. Construction best practice delivers greater cost and time certainty that comes from the actual out-turn cost never exceeding the forecast and the building always being completed on time.

Always demand evidence from design and construction firms to prove what has been achieved on other projects. Have the designers or the construction contractors measured the efficient utilisation of labour and materials through-out the site activities and can they provide you with the figures to back up their claims? Have they undertaken end-user surveys to test effective functionality and can they pro-vide you with copies of representative surveys? Have they compared out-turn costs with the initial budgets and can they tell you how often the two match? Have they compared the actual completion dates with the original target completion dates and can they tell you how often the two match? Have they any evidence from facility management staff of the maintainability, durability and ease of operation of com-pleted buildings and facilities?

It the results of this probing demonstrate that the design and construction firms are not able to prove they are con-sistently delivering the construction best practice of the six goals you would be wise to adopt the following action plan:

❑ Only invite expressions of interest from those firms that are able to provide well-documented evidence that they have practices and processes in place that ensure deliv-ery of the six primary goals. In particular, demand the names of those firms (including the design and other consultants, but particularly including specialist suppliers) that can prove they are already bound together within an integrated design and construction team in long-term strategic supply chain partnerships.

❑ If professional advice is deemed necessary, ensure those offering the advice can prove their deep and compre-hensive understanding of the six goals. They should also have a full understanding of the very comprehensive National Audit Office report *Modernising Construc-tion*. (The 'Be' organisation at Reading University and the 'Construction Best Practice Programme' at the Building Research Establishment can both assist in this area.)

❑ Ensure that you restrain your own organisation to defin-ing the business needs in detailed functional terms that are suitable for measuring and evaluating the output performance of the business activities housed within the built solution. Do not be tempted to define your requirement in terms of built solutions because to do so will transfer risk to yourself if it does not perform as efficiently as it should.

Self-Assessment Questionnaire for use in the Internal Change Process by Best Practice Clients

An effective change process must start by establishing accurately how well your current procurement process compares with the best practice to which you aspire. The following questions have been devised from my Building Down Barriers experience, with refinements from the St Helens Metropolitan Borough Council best value experience, and are offered to assist the self-assessment process. When answering each question against an individual goal, use the available evidence to assess the degree to which the criteria is met i.e. 0%, 10%, 20%, 30% of the time etc. Then calculate the average percentage for the specific goal, to show the gap in performance that your improvement process must close. Analysing the answers will show where the strengths and weaknesses of your current procurement process lie and will enable your change process to be targeted at the weakest areas of performance.

Functionality

❑ How frequently do the design and construction teams use value management workshops (that include end-users) to define and prioritise the detailed functional or business needs?

❑ How frequently are structured and facilitated value engineering workshops used by the design and construction teams to support and validate design decisions and the selection of all the components and materials?

❑ How many of those involved in the procurement process understand how maximum functionality would benefit the effectiveness and morale of end-users?

Cost of ownership

❑ How many of those involved in the procurement process have read the Confederation of Construction Clients' publication *Whole Life Costing: a client's guide?*

❑ How many of those involved in the procurement process have read the *Component Life Manuals* produced by the Building Performance Group (which is linked to the Housing Association Property Mutual insurance company)?

❑ How frequently do those involved in the procurement process demand that the design and construction team predict the cost of ownership?

❑ How well do those involved in the procurement process understand the difference between predicting the cost of ownership and estimating the cost of ownership?

❑ To what degree are those involved in the procurement process concerned about defects during the usage of a building and do they understand how these could affect the end-user's effectiveness?

Inefficiency and waste

❑ How many of those involved in the procurement process have any experience of working with construction teams that have measured the effective utilisation of labour and materials?

❑ How many of those involved in the procurement process have read the National Audit Office report *Modernising Construction,* particularly what it had to say about the degree to which the industry measures the effective utilisation of labour and materials?

❑ How many of those involved in the procurement process understand why measuring the effective utilisation of labour and materials is key to the elimination of unnecessary costs?

Specialist suppliers

❏ How many of those involved in the procurement process have direct experience of the significant savings that come directly from the early involvement of specialist suppliers in design?

❏ How many of those involved in the procurement process have read the Reading Construction Forum publication *Unlocking Specialist Potential*, or the Building Services Research and Information Association Technical Note TN14/97 *Improving M and E Site Productivity*?

❏ How many of those involved in the procurement process are committed to ensuring the early involvement of specialist suppliers in the design process in order to deliver 'right first time' on site?

Single point of contact

❏ How many of those involved in the procurement process have any direct experience of single point of contact procurement?

❏ How many of those involved in the procurement process believe that single point of contact is the only form of procurement that would ensure the involvement of specialist suppliers from the outset of design development and thus the delivery of 'right first time'?

Measurement

❏ How many of those involved in the procurement process understand why performance measurement is fundamental to the elimination of unnecessary costs?

❏ How many of those involved in the procurement process would agree with the DTI dictum that **'If you don't know how well you are doing, how do you know you are doing well?'** and are therefore demanding that construction contractors measure their effective utilisation of labour and materials?

To be a best practice client, the traditional assumption that selection of the design and construction team on the basis of lowest price will always ensure best value needs to be radically reappraised, since the assumption has rarely proved true in reality. There is overwhelming evidence to show that the tender price is invariably exceeded when the final account is presented; in many cases the final account has exceeded the tender price by over 60%. This fact was emphasised in the National Audit Office report *Modernising Construction* which stated:

'In 1999, a benchmarking study of 66 central government departments' construction projects with a total value of £500 million showed that three-quarters of the projects exceeded their budgets by up to 50% and two-thirds had exceeded their original completion date by 63%.'

The fact that the final settlement invariably exceeds the tender price by a considerable margin when the contract is awarded on the basis of the lowest price has been true for well over 100 years if John Ruskin is to be believed. He said in 1860 *'if you deal with the lowest bidder it is as well to add something for the risk you run, and if you do that, you have enough to pay for something better'*. The UK Prime Minister Margaret Thatcher said in the run-up to the 1979 election *'if you pay peanuts, you get monkeys'*.

Although she was talking about the civil service at the time and was emphasising the importance of paying salaries that would attract high calibre people, the expression was taken up by the wiser heads in the UK construction industry who have consistently argued that lowest price tendering rarely delivers value for money when the final settlement is rendered, and the picture becomes even worse when whole-life costs are taken into account.

It therefore follows from the above that for clients who are determined to get better value from procurement, especially better value in whole-life cost terms, selecting the design team separately from the construction team and the use of lowest price tendering must be replaced with some form of value-based selection that appoints the entirety of the design and construction supply chain from the outset on the basis of their skill and experience. This moves the selection process from a simplistic focus on lowest price to a much more sophisticated analysis of the skill and experience of each member of the design and construction supply chain. It also requires the introduction of a more open-book and trusting relationship between the client and the integrated supply-side team, which is founded on performance measurement and a partnering philosophy.

In the UK, government policy is that procurement should be on the basis of value for money and not lowest cost. The UK Minister of State for Trade and Industry has stated:

'The obsession with getting the lowest price for construction projects wastes money and cheats communities. Lowest cost does not mean better value.'

The more enlightened repeat clients in the UK have recognised that lowest price tendering has all too often turned out to be high risk to the commercial effectiveness of their business, because of the unpredictable price escalation that invariably occurs between the tender stage and the final settlement stage. They are exploring alternative ways of selecting an integrated supply-side team, and of awarding the contract, that place more reliance on trust, team working, open-book and the sharing of savings and risks. BAA, Argent and McDonald's are good examples of this in the UK private sector, and Defence Estates and St Helens Metropolitan Borough Council are good examples in the UK public sector.

The private sector clients that are moving away from tendering based on the lowest price tend to be major retailers who are using their skill and experience of managing their retail supply chain to actively manage their design and construction supply chain through various forms of supply chain partnerships. Whilst this approach by major retailers is both valid and low risk for clients that possess a high level of supply chain management skill and experience from their retail or manufacturing side, it may be high risk and difficult for clients that lack this skill or cannot afford the skilled resources that would be necessary to actively manage the design and construction supply chain.

A reality of construction procurement is that in almost every case, the client and the end user have a budget for the construction works that represents the maximum affordable price for constructing a building or facility to house the relevant functional or business activities. This may come from the business planning process that is an essential part of assessing the affordability of the venture in a competitive market, or it may simply come from the amount of money available from savings, grants or a loan.

In either case, if the final capital cost exceeds the maximum affordable price, the client will suffer some form of financial hardship and in the worst case (especially if the difference is up to 50%, as the National Audit Office found it was in three out of four public sector projects) this may well cause serious damage to the client's business competitiveness. This is equally true of the long-term operational and maintenance costs; the end-user client cannot afford unpleasant and costly financial surprises at any stage in the operational life of the building or facility because these will inevitably have to be paid for out of their profit margin and will therefore adversely affect their commercial competitiveness.

To avoid the consequences of 'employing monkeys', which can involve serious budget overruns, poor whole-life quality in terms of durability and poor functionality, the client needs to select an integrated design and construction team that are able to prove they have the skill and experience to deliver a construction solution that efficiently satisfies the client's business or functional needs and does not exceed the client's maximum affordable price.

The starting point for any repeat client that wishes to ensure their procurement of construction is consistently delivering best value in capital cost, whole-life cost or functional terms, is to measure how well their current procurement process has performed. To do this they need to review the constructed outputs of their procurement process in terms of the final capital cost and the performance in use of the completed building or constructed facility.

This requires an objective assessment of past construction works contracts to compare final settlements with tender prices, e.g. How often did the final settlement match the tender price? By how much did the final settlement exceed the tender price in the worst case? What is the average escalation in cost between the tender price and the final settlement? How often have the construction team been required to measure the effective utilisation of labour and materials? What is the average level of the effective utilisation of labour and materials?

It also requires an objective assessment of unexpectedly early component or materials failures during the planned life of the building or facility, and an assessment of the impact on the commercial well-being of the end user of making good the failure. Finally, it requires an objective assessment of the functional performance of the building or constructed facility by carefully interrogating the end users.

The situation is obviously different for small and occasional clients, since they do not have a continuous stream of construction contracts on which they can assess the effectiveness of their procurement process. Nevertheless, they would be well advised to ask their professional advisors about the true performance of the procurement approach they recommend. Similarly, they would be well advised to ask any potential construction contractor (including management contractors and design and build contractors) about their effectiveness in delivering value for money.

Any consideration of the appropriate procurement approach ought to bear firmly in mind that the Latham and the Egan Reports both made clear that design must be fully integrated with construction if clients are to achieve better value from construction procurement. The Egan Report, in particular, emphasised that the route to better performance by the industry and thus to better value for clients, was for clients to buy their built products in the same way they bought their manufactured products. This very clearly requires design to be an intrinsic part of the supply-side and to be totally integrated with construction using appropriate supply chain management techniques. Buying a built product in the same way as a manufactured product and use by the supply-side of supply chain management techniques also requires the supply-side to deal with the client through a single point of contact.

Whilst every demand-side client is free to develop its own value-based selection system, the following is suggested as a possible approach to the selection and appointment of the fully integrated design and construction team.

VALUE-BASED SELECTION OF A FULLY INTEGRATED DESIGN AND CONSTRUCTION TEAM – FOR USE BY ALL DEMAND-SIDE CLIENTS

Value-based selection differs radically from lowest price tendering, in that it either sets aside price completely or it constrains price to less than 20% of the available marks. The focus of value-based selection is on the evidence provided by the supply-side team of their skill, knowledge and experience. It is not about them merely telling you what they have done, it is about them showing you the evidence to prove they have done what they claim to have done.

The best supply-side team will be able to provide evidence of how effective their improvement process has been in driving out unnecessary costs caused by the ineffective utilisation of labour and materials. It will be able to provide evidence of the effectiveness of their value management, value engineering and risk management techniques on previous projects to deliver maximum functionality and the lowest optimum whole-life cost. It will be able to name those of its suppliers (sub-contractors, trades contractors and manufacturers) already involved in supply-side partnerships and will be able to give you the results of regular performance measurement from the continuous improvement targets that have been tied into the partnerships.

The suggested way forward in the introduction of value-based selection of the supply-side team is for the demand-side client to focus very firmly on the skill and experience of the supply-side team through the use of a carefully structured questionnaire. Since this is contrary to normal industry experience, it may be sensible for the demand-side client to emphasise to the industry that price is not part (or is not a significant part) of the selection process.

Ideally, the demand-side client should declare from the outset the maximum affordable price for the building or facility and should also make clear that the selected supply-side team will be required to develop a target cost that is within the maximum affordable price as the design is developed. If possible, the maximum affordable price (or gross maximum price) should be expressed in both initial capital cost and long-term operational cost terms, as the *Charter Handbook* emphasises that best practice clients are far more interested in best whole-life value.

Declaring the maximum affordable price (maximum gross price) from the outset and then setting aside the development of the target price until the supply-side design and construction team have been appointed, will lead to a far more accurate target price because it will be based on the completed design.

It is extremely difficult, if not impossible, for the supply-side design and construction team to establish an accurate target price for a specific building or facility before the design is fully developed and the ground conditions have been fully explored and converted into the actual, project-specific foundations. Without the developed design, all that can be deduced is a ballpark price based on data from similar buildings or facilities that have been constructed in recent years.

Unfortunately, such ballpark prices have to be heavily qualified (or loaded with contingency sums) because of the large number of unknowns (such as the ground conditions, the fabric of the building, the configuration and size of the building, the Town Planning constraints). Leaving the development of the target price until after the design has been developed eliminates the risk of price escalation caused by the unknowns. These unknowns are the things that regularly cause the price to escalate considerably during the design and construction process.

The experience of demand-side clients, such as St Helens Metropolitan Borough Council, that have opted for value-based selection of the supply-side team is that the supply-side team have always been able to establish a target price that was well within the maximum affordable price (gross maximum price).

Interestingly, I am also aware of three projects where the supply-side team was able to persuade the demand-side client to adjust the maximum affordable price upwards slightly because the use of through-life costs in the value engineering process had clearly demonstrated that the use of better quality and more durable components could pay back the higher capital cost within a few years. This decision was considerably helped by the integration of the end-user client into the supply-side team, because the client could immediately appreciate the cost of ownership benefit that would come from the use of the more durable components and could then quickly adjust their long-term business plan to check the affordability of the capital cost/operating cost adjustment.

Any demand-side client that is worried about the supposed risks of value-based selection would be well advised to make contact with those demand-side clients that have already embraced value-based selection and can therefore offer hard evidence of the results in terms of price and quality. Several local authorities and housing associations have already embraced value-based selection and are convinced from their experiences that it offers far better value for money than the traditional lowest price tendering. Several major private sector clients have also embraced a similar value-based selection approach and would be able to provide very convincing evidence to prove that value-based selection is better value than lowest price tendering. In the case of private sector clients, the simplest route to locate them would be through Be (previously the Design Build Foundation).

There are obviously other approaches to selecting the supply-side team by value that are equally valid, but no matter what approach is used, the client should always ensure that the process is based on the selection of a fully integrated design and construction team since this was advocated by the Egan Report and by the three standards (*Charter Handbook*, *Modernising Construction* and *Better Public Buildings*).

If price must be part of the value-based selection process, it should never count for more than 20% of the available marks and a two-envelope approach should be used so that the price can be dealt with separately from the skill and experience of the integrated team. It would also be wise to assess and award marks for the skills, knowledge and experience of the competing supply-side teams before price is considered, since there is always the risk that the prices in the second envelopes might colour the marking of the skills element of the bids.

Ideally, the envelopes containing the prices should not be opened until the contents of the skills envelopes have been assessed and marked. The delivery of best value will always come from the selection of the best design and construction team (in particular the selection of the best specialist suppliers), thus the marking system used should award the majority of the available marks to the skill and experience of the integrated team and should always do so on the basis of the well-documented evidence of performance that each team submits.

As I said above, the evidence submitted by the integrated design and construction supply-side team must include their recent achievements in delivering best value and at the very least ought to include the following:

- ❑ How often have final settlements matched tender prices?
- ❑ Where final settlements have exceeded tender prices, what was the worst case?
- ❑ What has been the average escalation in cost between tender price and final settlement?
- ❑ How often is the effective utilisation of labour and materials measured?
- ❑ What is the average level of the effective utilisation of labour and materials in recent construction works?
- ❑ What evidence have they of the true durability (performance in use) of the components and materials used in completed buildings or facilities?
- ❑ How often have they provided an accurate prediction of the annual cost of ownership of a building or facility?
- ❑ What evidence have they of end-user satisfaction with the functional performance of their completed buildings or facilities?

The above questions, which relate to the delivery of best value for other clients, would provide a minimum assurance that the selected team have a good track record in the necessary skills needed to deliver best value on the current contract.

However, where the achievement of best value is a commercial priority for the client, it might be wise to require the supply-side team to respond to a much more probing set of questions that could give greater assurance of their ability to deliver best value in terms of whole-life performance and the elimination of unnecessary costs. Basing the questionnaire on the six primary goals of construction best practice would ensure that all the necessary skills were adequately covered and would bring a degree of uniformity to the selection process. Such a questionnaire is suggested below.

VALUE-BASED SELECTION QUESTIONNAIRE FOR ASSESSING THE SKILL AND EXPERIENCE OF AN INTEGRATED DESIGN AND CONSTRUCTION TEAM

In order to bring a degree of commonality across the industry to that part of the selection process that evaluates skill and experience of the entire supply-side design and construction team, it would make sense to include a series of questions that relate to the six primary goals of construction best practice from the Construction Best Practice Programme booklet *A Guide to Best Practice in Construction Procurement* (these are also the six primary themes of construction best practice from the Rethinking Construction publication *Rethinking the Construction Client – Guidelines for construction clients in the public sector*).

It would also be essential to make the selection process as fair and objective as possible by basing the marking system on tangible evidence of experience, practice or performance rather than anecdotal claims. The following questionnaire may help the development of a value-based selection system, although it may need to be adjusted depending on the nature or size of the individual construction works.

When dealing with the response to a question against an individual goal, use the available evidence to assess the degree by which the criteria is met by all those involved in the design and construction process, i.e. 0%, 10%, 20%, 30% of the time, etc. When the responses to the individual questions have been assessed, take the average of the percentage awarded to each question to give the overall percentage compliance with the individual goal. When each goal has been dealt with in this manner, the assessment will show where the strengths and weaknesses of the competing design and construction teams lie and will enable the selection decision to be fair and objective.

Where the results of the value-based selection will have to stand up to internal (or external) audit, it would be sensible to consult closely with the auditors when devising the assessment questionnaire and to get their agreement to its final form. It should always be borne in mind that the unsuccessful bidders ought to have a right to be debriefed on the results of the assessment so that they can learn from the experience and understand what they need to do better next time.

A possible approach might be to issue the questionnaire to each competing design and construction team and ask them to submit a report that provides the response to each question, with examples of the evidence that supports the response. For instance, in the case of the first question about value management workshops against 'Functionality', they might state that the frequency was 20% of all projects over the last 5 years and include one or two value management workshop reports from representative projects with attendance lists for the workshops that include end users, designers, construction contractors and specialist suppliers. In the case of the first question about cost of ownership predictions against 'Cost of Ownership', they might state that the frequency was 5% of all projects over the last 5 years and include one or two cost of ownership predictions from representative projects.

It is essential to make clear to all the supply-side design and construction teams that are asked to make submissions that the term 'design and construction team' in the questionnaire refers to the entire supply chain and must therefore include the specialist suppliers (sub-contractors, trades contractors and manufacturers) that will actually be carrying out the construction activities for the building or facility.

Functionality
❑ How frequently are structured and facilitated value management workshops, that include end users, designers,

construction contractors and specialist suppliers, used to define and prioritise the detailed functional or business needs?

❏ How frequently are structured and facilitated value engineering workshops used by the design and construction teams to support and validate all design decisions and the selection of all the components and materials?

❏ How frequently are end users brought into the integrated teams and involved in value engineering workshops?

❏ How many of those involved in the design and construction team have received appropriate training to assist their understanding of the principles of the *Charter Handbook* and of the full implications of supply chain management?

❏ How frequently have end-user surveys been used after occupation to test the delight of the end users with the performance of the completed building?

Cost of ownership

❏ How frequently has the end-user client been provided with a cost of ownership prediction?

❏ Are such predictions always developed using relevant guidance, such as the Confederation of Construction Clients' publication *Whole Life Costing – A Client's Guide* and the *Component Life Manuals* produced by the Building Performance Group (which is linked to the Housing Association Property Mutual insurance company)?

❏ Can you provide evidence to prove that the design and construction team have received relevant training in predicting the cost of ownership?

Inefficiency and waste

❏ How frequently have the firms making up the design and construction team measured the effective utilisation of labour and materials?

❑ Can the firms making up the design and construction team provide evidence of relevant training in the measurement and elimination of the unnecessary costs that are generated by the ineffective utilisation of labour and materials?

❑ How do the firms making up the design and construction team (especially the specialist suppliers or trades contractors) measure improvements in productivity (their effective utilisation of labour and materials)?

Specialist suppliers

❑ How often are specialist suppliers (sub-contractors, trades contractors and manufacturers) involved in design from the outset?

❑ Have all the specialist suppliers in the design and construction team (sub-contractors, trades contractors and manufacturers) been selected using value-based selection rather than lowest price?

❑ Are all the members of the supply-side design and construction team tied together through supply-side partnering relationships?

❑ Can the construction contractor member of the supply-side team demonstrate a commitment to strategic long-term partnering with suppliers (sub-contractors, trades contractors and manufacturers)?

Single point of contact

❑ What proportion of the supply-side design and construction team has worked together on previous projects as an integrated supply-side team?

❑ Can all the firms in the supply-side team demonstrate a commitment to long-term supply-side partnering?

Measurement

❑ Do all the firms in the supply-side team benchmark their performance against Construction Best Practice

Programme Key Performance Indicators or other internal Key Performance Indicators?

☐ How often do the specialist suppliers in the supply-side team (sub-contractors, trades contractors and manufacturers) regularly record the actual man-hours worked (including abortive time) and the actual materials used (including wastage) and compare these with the forecast figures?

☐ How often do the firms in the design and construction supply-side team conduct a detailed analysis of abortive time and materials wastage, assess the causes and work with other team members to seek ways of eliminating those causes?

☐ Do the firms in the design and construction supply-side team set improvement targets that are aimed at reducing (and eventually eliminating) the gap between the forecast labour and materials figures and the actual labour and materials figures?

☐ Do the firms in the design and construction supply-side team open their books to other team members and share detailed performance data, such as profits, overheads and the effective utilisation of labour and materials?

Finally, remember that unless the demand-side clients change their procurement practices to a form that encourages and facilitates integration of the design and construction supply chain and the delivery of the highest optimum whole-life performance for the lowest optimum whole-life cost, the current drive for improvement will quickly die away.

Above all, do not forget the insistence in the *Accelerating Change* report that stated:

'Clients should require the use of integrated teams and long-term supply chains and actively participate in their creation.'

Further Reading and Help

For a short and definitive, plain-English explanation of best practice procurement:

A Guide to Best Practice in Construction Procurement

Available from the Construction Best Practice Programme, PO Box 147, Watford, WD25 9UZ. Telephone: 0845 605 55 56. Email: helpdesk@cbpp.org.uk Website: www.cbpp.org.uk

A simple guide written by the author of this book and aimed at staff 'at the sharp end' in all sectors of the industry, from end users to manufacturers. It explains the historical background to the Rethinking Construction movement and briefly describes the key aspects of the three best practice standards (*Better Public Buildings, Charter Handbook* and *Modernising Construction*). It then uses the analysis to list the six goals of construction procurement best practice (these were subsequently adopted by the UK Rethinking Construction organisation as the six themes of construction best practice). It also sets out the next steps that each sector must take in order to achieve the best practice of the three standards and warns of the consequences of doing nothing.

For a detailed, plain-English guide to best practice in design and construction:

Building Down Barriers – A guide to construction best practice

ISBN 0-415-28965-3
Available from Spon Press, 11 New Fetter Lane, London EC4P 4EE or 29 West 35th Street, New York, NY 10001.

A comprehensive and detailed guide written by the author of this book and aimed at staff 'at the sharp end' in all sectors of the industry, from end users to manufacturers, and developed from the concise Construction Best Practice Programme booklet *A Guide to Best Practice in Construction Procurement.* This book explains and compares, simply and clearly, the main aspects of the three UK best practice standards *(Better Public Buildings, Charter Handbook and Modernising Construction)*, compares them with similar developments in countries such as the USA and Singapore and lists the six goals of construction best practice that satisfy them all (these were adopted by the UK Rethinking Construction organisation as the six themes of construction best practice). It goes on to explain in detail the fundamental culture changes they necessitate in each sector of the UK construction industry and provides specific action plans for each sector that should deliver those cultural changes. The book is aimed at all members of the design and construction supply chain, including the client and end users, and it is also a valuable guide for students in all design and construction courses.

For a simple guide to self-assessment using the EFQM Business Excellence Model:

The Construction Performance Driver – A health check for your business
ISBN 1-90216-912-3
Available from BQC Performance Management Ltd, PO Box 175, Ipswich, IP2 8SW. Tel: 01473 409962. Fax: 01473 409966. Email: helpdesk@bqc-network.com Website: www.bqc-network.com

The guide provides an easy-to-use assessment tool, written specifically for the construction industry with considerable input from those at the sharp end within construction industry firms and based on the model of business excellence that has been widely used in major European and UK organisations since the early 1990s. The guide has been developed for use by all organisations operating within the construction industry (large or small, public or private) and who have a desire to improve their current business performance.

For the key UK reports advising on best practice in construction procurement:

Better Public Buildings
Available from the Department of Culture, Media and Sport, 2–4 Cockspur Street, London, SW1Y 5DH. Tel: 020 7211 6200. Website: www.culture.gov.uk/pdf/architecture.pdf

A short, lucid guide (six pages of text) that focuses strongly on the business benefits of well-designed buildings that enhance the quality of life, and therefore the efficiency, of the end users. It also explains the business benefits of using whole-life costs as the basis of design and construction decisions and it makes clear that best practice necessitates the appointment of integrated design and construction teams.

The Clients' Charter Handbook
Available from the Confederation of Construction Clients, 1st Floor, Maple House, 149 Tottenham Court Road, London, W1T 7NF. Tel: 020 7554 5340. Fax: 020 7554 5345. Email: cccreception@ccc-uk.co.uk Website: www.clientsuccess.org

A short, lucid guide (12 pages of text) that explains the approach to construction procurement that every chartered client must adopt. It focuses strongly on the importance of the client's leadership role within an integrated design and construction supply chain, which targets major reductions in whole-life costs, substantial improvements in functional efficiency and the elimination of defects over the whole life of the building. It also emphasises the benefits to repeat clients of long-term, partnering relationships with all key suppliers.

Modernising Construction
ISBN 0-10-276901-X
A report by the Comptroller and Auditor General of the National Audit Office (NAO) and available from any Stationary Office bookshop or by contacting the NAO. Tel: 020 7798

7400. Email: enquiries@nao.gsi.gov.uk Website: www.nao. gov.uk/publications/

A comprehensive report which sets out in detail the many barriers to improving construction industry performance and describes the various industry initiatives since 1994. It concludes that better value means better whole-life performance and that this can only come from total integration of the design and construction supply chain through a single point of contact. This ensures the involvement of the specialist suppliers in design from the outset, which is key to the elimination of inefficiency and waste, the achievement of optimum whole-life costs and the delivery of maximum functionality.

Local Government Task Force (LGTF) Rethinking Construction Toolkit
Available from the Customer Sales Department, Thomas Telford Ltd, Unit 1/K Paddock Wood Distribution Centre, Paddock Wood, Kent, TN12 6UU. Tel: 020 7665 2464. Fax: 020 7665 2245. Email: orders@thomastelford.com

This provides local authorities with a valuable support to the abandonment of outdated procurement practices that cause waste, in terms of the inefficient use of labour and materials, poor whole-life performance and poor functionality. It provides simple, practical 'how to' guidance that will enable local authority staff to introduce 'smart' procurement as recommended by the Egan Report.

For detailed guidance on supply chain integration and management:

The Building Down Barriers Handbook of Supply Chain Management – 'The Essentials'
ISBN 0-86017-546-4
Available from the Construction Industry Research and Information Association (CIRIA), 6 Storey's Gate, Westminster, London, SW1P 3AU.

An overview of the Building Down Barriers approach to supply chain integration and an introduction to the toolset as a whole. It describes the seven underlying principles of total supply chain integration and the lessons learned from their application on the two test-bed pilot projects. It also describes the benefits and the challenges of supply chain integration for the various sectors of the industry.

For guidance on predicting and validating whole-life costs:

Whole Life Costing – A Client's Guide
Available from the Confederation of Construction Clients, 1st Floor, Maple House, 149 Tottenham Court Road, London, W1T 7NF. Tel: 020 7554 5340. Fax: 020 7554 5345. Email: cccreception@ccc-uk.co.uk Website: www.client-success.org

A short, lucid guide (nine pages of text) that explains to clients the benefits, in business planning terms, of making construction investment decisions on the predicted cost of ownership. It explains the level of accuracy that can be expected at the various stages of design and construction and makes clear that optimum whole-life costs can only be achieved with the early involvement of specialist suppliers in design.

Technical Audit of Building and Component Methodology
Available from the Building Performance Group, Grosvenor House, 141–143 Drury Lane, London, WC2B 5TS. Tel: 020 7240 8070.

Describes a technical audit process for assessing the whole life performance of buildings and can be used as a first, second or third party audit system.

Housing Association Property Mutual (HAPM) Component Life Manual
Available from E and F N Spon, Cheriton House, North Way, Andover, Hampshire, SPIO 5BE. Tel: 01264 342933.

The Manual schedules over 500 components and gives the insured life, maintenance requirements and adjustment factors. The insured lives are cautious, were developed for housing, and are limited to 35 years, so should not be used without adjustment. The Manual is updated twice a year, contains references to current British and European Standards and includes feedback from research and claims on HAPM latent defect insurance.

Building Services Component Life Manual – Building Life-plans
Available from Blackwell Publishing, 9600 Garsington Rd, Oxford, OX4 2OQ. Tel: 01865 776868. Fax 01865 714591.

   This Manual provides much needed guidance on the longevity and maintenance requirements of mechanical and electrical plant. It sets out typical lifespans of building service components – boilers, pipes, ventilating systems, hydraulic lifts, etc. These are ranked according to recognised benchmarks of specification, together with adjustment factors for differing environments, use patterns and operating regimes. Summaries of typical inspection and maintenance requirements are provided, along with specification guidance and references to further sources of information.

For detailed guidance on involving specialist suppliers (trade and specialist contractors) in design:

Unlocking Specialist Potential
ISBN 1-902266-00-5
Available from Be (previously Reading Construction Forum), PO Box 2874, London Road, Reading, Berkshire, RG1 5UQ Tel: 0118 931 8190.

   A detailed guide that explains how the skill and experience of specialist suppliers can be harnessed in design development. It proposes strategies for better teamwork and collaboration, for a process orientated approach to design and construction, and for a central focus on customer requirements. It makes clear that it is only by enabling the specialist suppliers to play a key role within the design process that real improvements in value can be achieved. The guide was used to develop the technology cluster concept in the Building Down Barriers toolset.

For detailed international evidence of labour inefficiency levels:

BSRIA Technical Note 14/97, Improving M & E Site Productivity

Available from the Building Services Research and Information Association, Old Bracknell Lane West, Bracknell, Berkshire, RG12 7AH. Tel: 01344 426511. Fax: 01344 487575. Email: bookshop@bsria.co.uk Website: www.bsria.co.uk

Comprehensive evidence from projects in the UK, USA, Germany, France and Sweden on the true level of the efficient use of labour in mechanical and electrical services. It also describes the causes of the inefficiencies described in the report, including naming the sector of the industry that was responsible for causing the individual problem. It also gives advice on how to improve efficiency levels by better integration and co-ordination.

BSRIA Technical Note 13/2002, Site Productivity – 2002, A guide to the uptake of improvements

Available from the Building Services Research and Information Association, Old Bracknell Lane West, Bracknell, Berkshire, RG12 7AH. Tel: 01344 426511. Fax: 01344 487575. Email: bookshop@bsria.co.uk Website: www.bsria.co.uk

A comprehensive review of what has happened in the UK building services industry since the publication of TN 14/97, including detailed feedback from the small number of projects where the recommendations from TN 14/97 have been applied.

For an analysis of the key reports on the UK construction industry since 1944:

Construction Reports 1944-98

ISBN 0-632-05928-1
Available from Blackwell Publishing 9600 Garsington Rd, Oxford, OX4 2DQ. Tel: 01865 776868 Fax: 01865 714591

A detailed analysis of the key reports on the UK construction industry since 1944, starting with the Simon Committee report in 1944 and ending with the Egan report in 1998. In its conclusion it picks up the recurring themes that run through the reports and thus demonstrates how little the industry's structure, culture and performance has improved since 1944.

For measurement of effective labour utilisation:

CALIBRE The productivity toolkit

For further information contact the Centre for Performance Improvements in Construction (CPIC), BRE, Garston, Watford, Hertfordshire, WD2 7JR.

CALIBRE provides a consistent and reliable way of identifying how much time is being spent on activities that directly add value to the construction and how much time is being spent on non-added value activities.

For training and coaching in construction and procurement best practice:

ICOM/CITB Diploma in Construction Process Management

For further information contact ICOM, Long Grove House, Seer Green, Buckinghamshire, HP9 2UL. Tel: 01494 675921. Fax: 01494 675126. Email: Kinder.ICOM@btinternet.com

ICOM is linked with the Construction Industry Training Board (CITB) and the University of Cambridge Local Examinations Syndicate, and uses the Construction Best Practice Programme booklet *A Guide to Best Practice in Construction Procurement* to define best practice in its training. ICOM is also working with the CITB to offer awareness workshops and coaching for clients on best practice procurement. ICOM is also working with CITB to develop coaching and training in best practice construction for small and medium sized construction industry firms (design consultants, construction contractors, specialist/trades contractors and manufacturers).

For advice on finding and working with integrated design and construction teams:

Be (previously the Design Build Foundation and Reading Construction Forum)

PO Box 2874, London Road, Reading, RG1 5UQ. Tel: 0118 931 8190 Fax: 0118 975 0404 Email: enquiries@dbf.uk.com Website: www.dbf-web.co.uk

Be brings together representatives from the whole construction industry to champion the total integration of design and construction in order to deliver customer satisfaction through a single source of responsibility. It was formed from the merging of the Design Build Foundation and Reading Construction Forum in October 2002 and is a self-funded, multi-discipline organisation comprising leading construction industry clients, designers, consultants, contractors, specialists, manufacturers and advisors. Be is the UK's largest, independent organisation for companies across the whole supply chain.

Index